Springer Series in Geomechanics and Geoengineering

Series Editors

Wei Wu
Ronaldo I. Borja

For further volumes:
http://www.springer.com/8069

Serguey A. Elsoufiev

Strength Analysis in Geomechanics

Second Edition

Series Editors

Wei Wu
Ronaldo I. Borja

Springer

Professor Wei Wu, Institut für Geotechnik, Universität für Bodenkultur, Feistmantelstraße 4, 1180 Vienna, Austria, E-mail: wei.wu@boku.ac.at

Professor Ronaldo I. Borja, Department of Civil and Environmental Engineering, Stanford University, Stanford, CA 94305-4020, USA, E-mail: borja@stanford.edu

Prof. Dr. Serguey A. Elsoufiev
Vierhausstr. 27
44807 Bochum
Germany
elsoufie@et.rub.de

ISSN 1866-8755
ISBN 978-3-642-26180-0 e-ISBN 978-3-642-01301-0
DOI 10.1007/978-3-642-01301-0
Springer Dordrecht Heidelberg London New York

Cover design: Erich Kirchner, Heidelberg

Printed on acid-free paper

Springer is part of Springer Science+Business Media (www.springer.com)

Foreword

It is hardly possible to find a single rheological law for all the soils. However they have mechanical properties (elasticity, plasticity, creep, damage etc.) that are met in some special sciences, and basic equations of these disciplines can be applied to earth structures. This way is taken in this book. It represents the results that can be used as a base for computations in many fields of the Geomechanics in its wide sense. Deformation and fracture of many objects include a row of important effects that must be taken into account. Some of them can be considered in the rheological law that, however, must be simple enough to solve the problems for real objects.

On the base of experiments and some theoretical investigations the constitutive equations that take into account large strains, a non-linear unsteady creep, an influence of a stress state type, an initial anisotropy and a damage are introduced. The test results show that they can be used first of all to finding ultimate state of structures – for a wide variety of monotonous loadings when effective strain does not diminish, and include some interrupted, step-wise and even cycling changes of stresses. When the influence of time is negligible the basic expressions become the constitutive equations of the plasticity theory generalized here. At limit values of the exponent of a hardening law the last ones give the Hooke's and the Prandtl's diagrams. Together with the basic relations of continuum mechanics they are used to describe the deformation of many objects. Any its stage can be taken as maximum allowable one but it is more convenient to predict a failure according to the criterion of infinite strains rate at the beginning of unstable deformation. The method reveals the influence of the form and dimensions of the structure on its ultimate state that are not considered by classical approaches.

Certainly it is hardly possible to solve any real problem without some assumptions of geometrical type. Here the tasks are distinguished as anti-plane (longitudinal shear), plane and axisymmetric problems. This allows to consider a fracture of many real structures. The results are represented by relations that can be applied directly and a computer is used (if necessary) on a final stage of calculations. The method can be realized not only in

Geomechanics but in other branches of industry and science. The whole approach takes into account 5 types of non-linearity (3 physical and 2 geometrical) and contains some new ideas, for example, the consideration of the fracture as a process, the difference between the body and the element of a material which only deforms and fails because it is in a structure, the simplicity of some non-linear computations against linear ones (ideal plasticity versus the Hooke's law, unsteady creep instead of a steady one etc.), the independence of maximum critical strain for brittle materials on the types of structure and stress state, an advantage of deformation theories before flow ones and others.

All this does not deny the classical methods that are also used in the book which is addressed to students, scientists and engineers who are busy with strength problems.

Foreword to the 2nd Edition

More than 2 years passed from the appearance of my work "Strength Analysis in Geomechanics". Since that time I worked on the theme, solved some new geotechnical problems and included them in the book. Among them the fracture of a tunnel arch in Chap. 1 and of economical profile of triangular dam under self-weight as well as lateral pressure in Chap. 2; ultimate state of compressed soil by inclined plates in Chap. 4; displacements in massif pressed similarly in Chap. 5; generalization of shells theory and that of compressed cylinder in Chap. 6; practical approach to the task of plastic material flow in the cone (Appendix E); propagation of big plastic zones near crack adges in Appendix F; strength of different profiles rotating disk in Appendix O, ultimate state of anisotropic tubes under axial load and internal pressure; the same for isotropic ones with addition of torsion are found; construction of rheological equations based on the hypothesis of potential function existence; comparison with test data on metals and a polymer. Some important additions are made. Among them ultimte state of a non-linear material at eccentric compression in Chap. 1; direct equations for main (principal) stresses at a plane problem in Chap. 2; deformation of soil under triangular load, displacements of the beam on elastic foundation under uniformly distributed load on finite length and in slope of compressed body (Chap. 3); critical loads for constant displacements in Chaps. 3 and 5 and Appendixes F, I, J are found. New references are also attached. Some figures are corrected and slips of the pen as well as misprints are remowed. The changes above demand to correct the title of the book as "Strength Analysis in Geotechnics".

2009 *S. Elsoufiev*

Preface

The solution of complex problems of strength in many branches of industry and science is impossible without a knowledge of fracture processes. Last 50 years demonstrated a great interest to these problems that was stimulated by their immense practical importance. Exact methods of solution aimed at finding fields of stresses and strains based on theories of elasticity, plasticity, creep, etc. and a rough appreciation of strength provide different results and this discrepancy can be explained by the fact that the fracture is a complex problem at the intersection of physics of solids, mechanics of media and material sciences. Real materials contain many defects of different form and dimensions beginning from submicroscopic ones to big pores and main cracks. Because of that the use of physical theories for a quantitative appreciation of real structures can be considered by us as of little perspective. For technical applications the concept of fracture in terms of methods of continuum mechanics plays an important role. We shall distinguish between the strength of a material (considered as an element of it – a cube, for example) and that of structures, which include also samples (of a material) of a different kind. We shall also distinguish between various types of fracture: ductile (plastic at big residual strains), brittle (at small changes of a bodies' dimensions) and due to a development of main cracks (splits).

Here we will not use the usual approach to strength computation when the distribution of stresses are found by methods of continuum mechanics and then hypotheses of strength are applied to the most dangerous points. Instead, we consider the fracture as a process developing in time according to constitutive equations taking into account large strains of unsteady creep and damage (development of internal defects). Any stage of the structures' deformation can be supposed as a dangerous one and hence the condition of maximum allowable strains can be used. But more convenient is the application of a criterion of an infinite strain rate at the moment of beginning of unstable deformation. This approach gives critical strains and the time in a natural way. When the influence of the latter is small, ultimate loads may be also found. Now we show how this idea is applied to structures made of different materials, mainly soils.

The first (introductory) chapter begins with a description of the role of engineering geological investigations. It is underlined that foundations should not be considered separately from structures. Then the components of geomechanics are listed as well as the main tasks of the Soil Mechanics. Its short history is given. The description of soil properties by methods of mechanics is represented. The idea is introduced that the failure of a structure is a process, the study of which can describe its final stage. Among the examples of this are: stability of a ring under external pressure and of a bar under compression and torsion (they are represented as particular cases of the common approach to the stability of bars); elementary theory of crack propagation; the ultimate state of structures made of ideal plastic materials; the simplest theory of retaining wall; a long-time strength according to a criterion of infinite elongations and that of their rate. The properties of introduced non-linear equations for unsteady creep with damage as well as a method of determination of creep and fracture parameters from tests in tension, compression and bending are given (as particular cases of an eccentric compression of a bar).

In order to apply the methods of Chap. 1 to real objects we must introduce main equations for a complex stress state that is made in Chap. 2. The stresses and stress tensor are introduced. They are linked by three equilibrium equations and hence the problem is statically indeterminate. To solve the task, displacements and strains are introduced. The latters are linked by compatibility equations. The consideration of rheological laws begins with the Hooke's equations and their generalization for non-linear steady and unsteady creep is given. The last option includes a damage parameter. Then basic expressions for anisotropic materials are considered. The case of transversally isotropic plate is described in detail. It is shown that the great influence of anisotropy on rheology of the body in three options of isotropy and loading planes interposition takes place. Since the problem for general case cannot be solved even for simple bodies some geometrical hypotheses are introduced. For anti-plane deformation we have five equations for five unknowns and the task can be solved easily. The transition to polar coordinates is given. For a plane problem we have eight equations for eight unknowns. A very useful and unknown from the literature combination of static equations is received. The basic expressions for axi-symmetric problem are given. For spherical coordinates a useful combination of equilibrium laws is also derived.

It is not possible to give all the elastic solutions of geo-technical problems. They are widely represented in the literature. But some of them are included in Chap. 3 for an understanding of further non-linear results. We begin with longitudinal shear which, due to the use of complex variables, opens the way to solution of similar plane tasks. The convenience of the approach is based on the opportunity to apply a conformal transformation when the results for simple figures (circle or semi-plane) can be applied to compound sections. The displacement of a strip, deformation of a massif with a circular hole and a brittle rupture of a body with a crack are considered. The plane deformation of a wedge under an one-sided load, concentrated force in its apex and pressed by inclined plates is also studied. The use of complex variables is demonstrated on the task of compression of a massif with a circular hole. General relations for a semi-plane under a vertical load are applied to the cases of the crack in tension and a constant displacement under a punch. In a similar way relations for transversal shear are used, and critical stresses are found. Among the axi-symmetric problems a sphere, cylinder and cone under internal and external pressures are investigated. The generalization of the Boussinesq's problem includes determination of stresses and displacements under loads uniformly distributed in a circle and rectangle. Some approximate approaches for a computation of the settling are also considered. Among them the layer-by-layer summation and with the help of the so-called equivalent layer. Short information on bending of thin plates and their ultimate state is described. As a conclusion, relations for displacements and stresses caused by a circular crack in tension are given.

Many materials demonstrate at loading a yielding part of the stress-strain diagram and their ultimate state can be found according to the Prandtl's and the Coulomb's laws which are considered in Chap. 4, devoted to the ultimate state of elastic-plastic structures. The investigations and natural observations show that the method can be also applied to brittle fracture. This approach is simpler than the consequent elastic one, and many problems can be solved on the basis of static equations and the yielding condition, for example, the torsion problem, which is used for the determination of a shear strength of many materials including soils. The rigorous solutions for the problems of cracks and plastic zones near punch edges at longitudinal shear are given. Elastic-plastic deformation and failure of a slope under vertical loads are studied among the plane problems. The rigorous solution of a massif compressed by inclined plates for particular cases of soil pressure on a retaining wall and flow of the earth between two foundations is given. Engineering relations for wedge penetration and a load-bearing capacity of a piles sheet are also presented. The introduced theory of slip lines opens the way to finding the ultimate state of structures by a construction of plastic fields. The investigated penetration of the wedge gives in a particular case the ultimate load for punch pressure in a medium and that with a crack in tension. A similar procedure for soils is reduced to ultimate state of a slope and the second critical load on foundation. Interaction of a soil with a retaining wall, stability of footings and different methods of slope stability appreciation are also given. The ultimate state of thick-walled structures under internal and external pressures and compression of a cylinder by rough plates are considered among axi-symmetric problems. A solution to a problem of flow of a material within a cone, its penetration in a soil and load-bearing capacity of a circular pile are of a high practical value.

Many materials demonstrate a non-linear stress-strain behaviour from the beginning of a loading, which is accompanied as a rule by creep and damage. This case is studied in Chap. 5 devoted to the ultimate state of structures at small non-linear strains. The rigorous solution for propagation of a crack and plastic zones near punch edges at anti-plane deformation is given. The generalization of the Flamant's results and the analysis of them are presented. Deformation and fracture of a slope under vertical loads are considered in terms of simple engineering relations. The problem of a wedge pressed by inclined plates and a flow of a material between them as well as penetration of a wedge and load-bearing capacity of piles sheet are also discussed. The problem of the propagation of a crack and plastic zones near punch edges at tension and compression as well as at transversal shear are also studied. A load-bearing capacity of sliding supports is investigated. A generalization of the Boussinesq's problem and its practical analysis are fulfilled. The flow of the material within a cone, its penetration in a massif and the load-bearing capacity of a circular pile are studied. As a conclusion the fracture of thick-walled elements (an axi-symmetrically stretched plate with a hole, sphere, cylinder and cone under internal and external pressures) are investigated. The results of these solutions can be used to predict failure of the voids of different form and dimensions in soil.

In the first part of Chap. 6, devoted to the ultimate state of structures at finite strains, the Hoff's method of infinite elongations at the moment of fracture is used. A plate and a bar at tension under hydrostatic pressure are considered. Thick-walled elements (axi-symmetrically stretched plate with a hole, sphere, cylinder and cone under internal and external pressure) are studied in the same way. The reference to other structures is made. The second part of the chapter is devoted to mixed fracture at unsteady creep. The same problems from its first part are investigated and the comparison with the results by the Hoff's method is made. The ultimate state of shells (a cylinder and a torus of revolution) under internal pressure as well as different membranes under hydrostatic loading is studied. The comparison with test data is given. The same is made for a short bar in tension and compression. In conclusion the fracture of an anisotropic plate in biaxial tension is investigated. The results are important not only for similar structures but also for a finding the theoretical ultimate state of a material element (a cube), which are usually formulated according to the strength hypotheses. The found independence of critical maximum strain for brittle materials on the form of a structure and the stress state type can be formulated as a "law of nature".

Contents

1

Introduction: Main Ideas

1.1 Role of Engineering Geological Investigations

An estimation of conditions of buildings and structures installation demands a prediction of geological processes that can appear due to natural causes or as a result of human activity. This prediction should be based on a geological analysis which takes into account different forms of interaction between created structures and an environment.

The prediction of the structures role is usually made by the methods of the engineering geology that includes computations according to laws of mechanics (deformation, stability etc.). The basic data are: geological schemes and cross-sections, physical-mechanical characteristics of soils, and others. An engineer-geologist participates in the choice of structures places and gives their proofs.

The engineer-geologist must always take into account that engineering structures made of concrete, brick, rock, steel, wood and other materials should not be considered separately of their foundations which must have the same degree of reliability as a whole construction. The main causes of their destruction can be changes in their stress state and, consequently, – the deformation of the soil. So, an engineer and a geologist must have enough knowledge of the geo-mechanics as a part of the whole structures theory.

1.2 Scope and Aim of the Subject. Short History of Soil Mechanics

The Soil Mechanics deals mainly with sediments and their unconsolidated accumulations of solid particles produced by mechanical and chemical disintegration of rocks. It includes the parts (according to another classification they are components of the geo-mechanics): (1) Mechanics of rocks. (2) That of mantel or rigolitth – the Soil Mechanics itself. (3) Mechanics of organic masses. (4) Mechanics of frozen earth.

The main features of soils are:

a) A disintegration which distinguishes a soil from a rock, the particles can not be bonded but they form a body with the strength much lower than that of a particle; it induces a porosity that can change under external actions;
b) They have the property of permeability;
c) Strength and stability of a soil is a function of a cohesiveness and of a friction between the particles;
d) The stress-strain dependence of a soil includes as a rule residual components and influence of time (a creep phenomenon).

The main tasks of the Soil Mechanics are as follows: (1) the establishment of basic laws for soils as sediments and other accumulations of particles. (2) the study of soil strength and stability including their pressure on retaining walls. (3) investigation of structures strength problems in different phases of deformation.

For the solution of these tasks two main methods are used – theoretical (on the basis of a mathematical approach) and modelling with different materials. Here the first of them is considered.

The development of the Soil Mechanics began at the end of the eighteenth century. The first period is characterized by a rare use of scientific methods. The theoretical investigations that are actual now are contained in the works of French scientists (Prony, 1810; Coulomb, 1778; Belidor, 1729; Poncelet, 1830 and others) who were solving the problem of soil pressure on retaining wall and Russian academician Fussa (the end of the eighteenth century) who received a computational method for a beam on elastic foundation. Until the beginning of the twentieths century works on the Soil Mechanics were linked with determination of a soil pressure on retaining walls and a solution of the simplest problems of slopes as well as footings stability. V.I. Kurdumov began in 1889 laboratory tests of soils as foundations of structures.

The next step in the Soil Mechanics was made by Carl Terzaghi /1/ in USA and N.M. Gersewanov in the USSR. They gave the schemes of deformation calculations. In the problem of soil strength we must mention the works of N.N. Pouzyrevski in the USSR and O. Frölich /2/ in Germany. The books of N.A. Cytovich /3/, V.A. Florin, N.N. Maslov and other Russian scientists have broad applications. Rigorous solutions for a soil massif at its ultimate state was given by V.V. Sokoöovski /4/.

1.3 Use of the Continuum Mechanics Methods

Computations in the Soil Mechanics are usually fulfilled by methods of the Continuum Mechanics. Although soils have different mechanical properties their settlings have been found on the base of the Elasticity Theory /5/. The complete solution of its plane problem was given by N.I. Muschelisvili /6/

thanks to the use of the complex variables. An application of the Plasticity Theory methods are met much rarer although they give solutions nearer to the reality /7/.

The founder of the strength disciplines is G. Galilei who in 1638 published his book "Discorsi E Demonstrazioni Matematiche Intorno A Due Nuove Science" (Talks and proofs concerning two new sciences) in which he grounded the Theoretical Mechanics and the Strength of Materials. In that time laws of deformation were not discovered and G. Galilei appreciated a strength of bodies directly.

The discovery of the linear dependence between acting force and induced by it displacement by R. Hooke in 1676 gave the basis of the Elasticity Theory. Rigorous definitions of stresses and strains formulated a civil engineer O. Cauchy. The main mathematical apparatus of this science was introduced in the works of G. Lame and B. Clapeyron who worked at that period in S.-Petersburg institute of Transport Communications. The number of practically important problems was solved thanks to the famous principle of Saint-Venant.

In calculations according to the scheme of the Elasticity Theory the main task is the determination of stress and strain fields. An estimation of a strength has as a rule an auxiliary character since a destruction in one point or in a group of them does not lead to a failure of a structure. The Galilei's idea of an appreciation of the strength of the whole body found applications only in some districts of the Continuum Mechanics (the stability of compressed bars according to the Euler's approach, a failure of some objects in the Structural Mechanics, the theory of the ultimate state of soils, and quite recently – in the theory of cracks propagation due to the Griffith's idea /8/).

The computations according to the ultimate state began to develop thanks to a study of metal plastic deformation at the end of the nineteenth century by French investigators Levy, Tresca, Saint-Venant and at the beginning of the last century by German scientists Mises, Hencki, Prandtl. The latter introduced the diagram of an ideal elastic-plastic material and solved a row of important problems including geo-mechanic ones. The important role for the practice play up to now two Gvozdev's theorems of the ultimate state of a plastic body /9/ (the static one that proposes the ultimate load as a maximum force among all corresponding to an equilibrium and a minimum load for all kinematically possible forms of destruction) which he proved in the thirties of the last century in the USSR for the objects of the Structural Mechanics. In the Media Mechanics such theorems were formulated at the fifties in USA.

The prominent contribution to the Plasticity Theory made W. Prager, F. Hodge and A. Nadai who received also a row of important results in the geomechanics and other disciplines. The works of Soviet scientists V. Sokolovski, L. Kachanov and A. Iljushin in this field are also well-known. The special interest have their investigations in the Theory of Plasticity of a hardening material which describes the real behaviour of a continuum and includes as particular cases the linear elasticity and the ideal plasticity. An intensive study of cracks in an elastic-plastic material provides G. Rice in USA.

The phenomenon of creep was discovered by physicists who (Boltzman, Maxwell, Voight, Kelvin) constructed in the nineteenth century the constitutive equations which are actual now. In the technique the creep processes are studied since the twentieths of the last century in the connection with the metals deformation at elevated temperatures under constant loads. The construction of the basic relations followed the ideas of the Plasticity Theory of a hardening body. The large work in this direction was made by F. Odquist.

In twentieths-thirtieths of the last century Odquist and Hencki found an opportunity to compute the fracture time of a bar in tension under constant load when its elongation tends to infinity. This idea began to spread out only after the work of N. Hoff (1953) who used more simple equation of creep and received the good agreement with test data. To predict an earlier failure of the structure L.M. Kachanov introduced in the sixtieths a parameter of a damage as a ratio of a destructed part of a cross-section to the whole one. According to his idea the bar either elongates infinitely or is divided in parts when the defects fit up the whole area.

Another way to describe the ultimate state in a creep opens the criterion of infinite elongations rate at the beginning of unstable deformation (R. Carlsson, 1966). The introduction in constitutive equations of a damage parameter allowed S. Elsoufiev to find on the base of the criterion the ultimate state of many objects including geotechnical ones /7.10/.

1.4 Main Properties of Soils

1.4.1 Stresses in Soil

Due to the weight (which is always present), tectonic, hydrodynamic, physical-chemical, residual and other processes internal stresses appear in the earth. In a weightless massif at an action of load P (Fig. 1.1) in a point M a part of the body under cross-section nMl is in an equilibrium with internal stresses p which are distributed non-uniformly in the part of the massif. If they are constant in a cross-section the relation for their determination (Fig. 1.2) is

$$\mathrm{p} = \mathrm{P}/\mathrm{A} \tag{1.1}$$

where A is an area of cross-section aa.

At a non-uniformly distributed stresses they can be found according to expression

$$\mathrm{p} = \lim_{\mathrm{dA}\to 0}(\mathrm{dP}/\mathrm{dA}). \tag{1.2}$$

Here dA is an elementary area in a surrounding of the investigated point and dP – the resultant of forces acting in it.

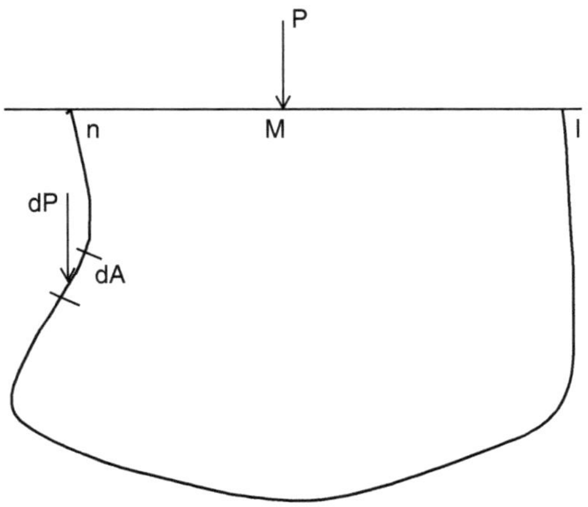

Fig. 1.1. Stresses in massif of soil

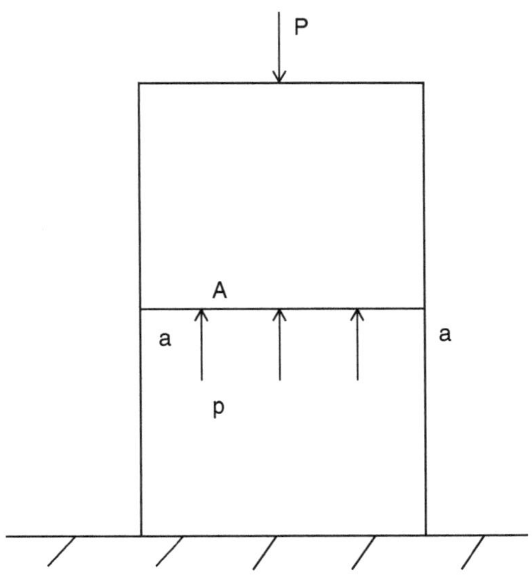

Fig. 1.2. Stresses in compressed bar

The value and the direction of stress p depend not only on a meaning of external forces and the position of the point but also on the direction of a cross-section. If vector p is inclined to a plane it can be decomposed into normal σ and shearing τ components (Fig. 1.3).

Since materials resist differently to their actions such a decomposition has a physical meaning. In the general case an elementary cube is cut around the point on each side of which one normal and two shearing stresses act (see Chap. 2 further).

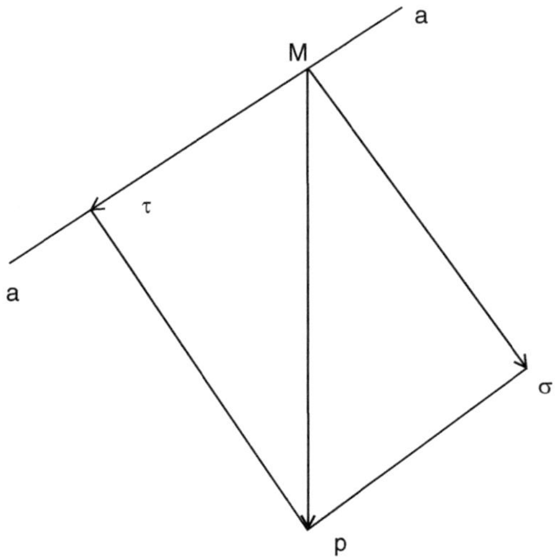

Fig. 1.3. Decomposition of stress p

The base of a structure is a part of the massif where stresses depend on the structure erected. It differs from a foundation that transfers the structure weight to the base. The boundary of the latter is a surface where the stresses are negligible. All the artificial soil massifs (embankments, dams etc.) are not the bases, they are structures.

Stresses in the massif under external loads differ from their real meanings on values of the soil self-weight components. These so-called natural pressures depend on a specific weight γ_e of the soil, a coordinate z of the point and the depth of an underground water. The natural pressure is determined by relation

$$\sigma_z' = \gamma_e z + \gamma' z'$$

where γ' is the specific weight of the soil with the consideration of its suspension by the underground waters, z' is the depth of the point from their mirror. Normal stresses on vertical planes are determined as

$$\sigma_x' = \sigma_y' = \zeta \sigma_z'.$$

Here $\zeta = \nu/(1-\nu)$ is the factor of lateral soil expansion and ν – the Poisson's ratio.

1.4.2 Settling of Soil

Phases of Soil State

An elastic solution shows that with a growth of a load on a punch plastic deformation begins at its edges. Then inelastic zones expand according to the plastic solution. Experiments confirm this picture if we suppose that the

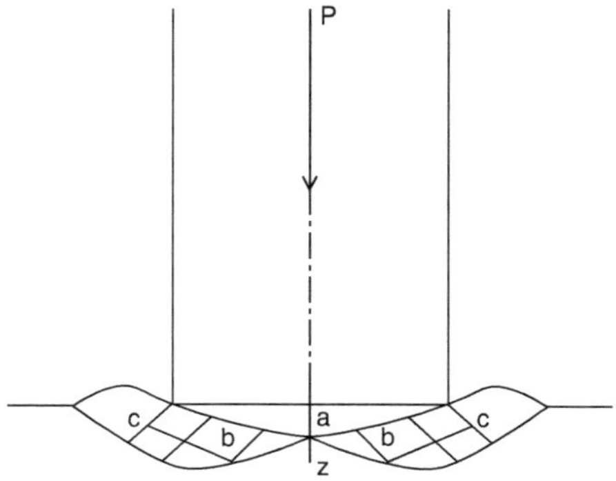

Fig. 1.4. Zones of stress state under punch

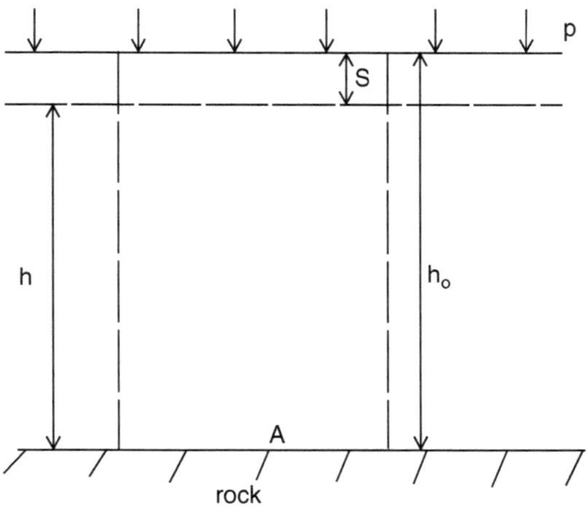

Fig. 1.5. Settling of a layer of limited thickness

part of the soil (districts a, b in Fig. 1.4) acts together with the punch. At the same time the expansion of the soil upwards (zones c in the figure) takes place and slip lines appear in zones b. This phenomenon was observed by V.I. Kourdumov in his tests. Professor N.M. Gersewanov proposed to consider three stages of the base state at the growth of the load: (1) its condensation, (2) an appearance of the shearing displacements and (3) its expansion. By these processes a condensed solid core (zone a in the figure takes place) and it moves together with the punch making additional plastic districts.

Settling of Earth Layer of Limited Thickness

Under an action of uniformly distributed load p at a large length (Fig. 1.5) an earth layer is exposed to a pressure without lateral expansion. The process is similar to the compressive deformation and the problem becomes one-dimensional. If the layer is supported by an incompressible and impenetrable basis its full settling is equal to the difference of current and initial lengths, that is

$$S = h_o - h \tag{1.3}$$

The skeleton volume in a prism with a basic area A before and after the deformation remains constant as

$$Ah_o/(1 + e_o) = Ah/(1 + e) \tag{1.4}$$

where e_o, e are factors of the soil porosity before and after the loading. They are computed as ratios of pores and skeleton volumes.

Solving (1.4) relatively to height h and putting it into (1.3) we find

$$S = h_o(e_o - e)/(1 + e_o). \tag{1.5}$$

Now we introduce a factor a as

$$a = (e_o - e)/p$$

and put it in (1.5) which gives

$$S = h_o ap/(1 + e_o). \tag{1.6}$$

Value

$$a/(1 + e_o)$$

is a factor a_v of soil compressibility and (1.6) becomes

$$S = h_o a_v p. \tag{1.7}$$

As a result we receive that the full settling of the soil layer under a homogeneous loading and in the conditions of an absence of its lateral expansion is proportional to the thickness of the layer, the intensity of the load and depends on the properties of the soil.

Role of Loading Area

As natural observations show a settling depends on the loading area in a form of the curve in Fig. 1.6 on which three districts can be distinguished: 1 – of small areas (till $0.25\,m^2$) when the soil is in the phase of the shearing displacements and the settling decreases with a growth of A, a zone 2 where the soil is in the phase of a condensation and the settling is practically proportional to $A^{0.5}$:

$$S = Cp\sqrt{A}$$

where C is a coefficient of proportionality, and part 3 in which with a fall of the condensation role of the kern the divergences from the proportionality law

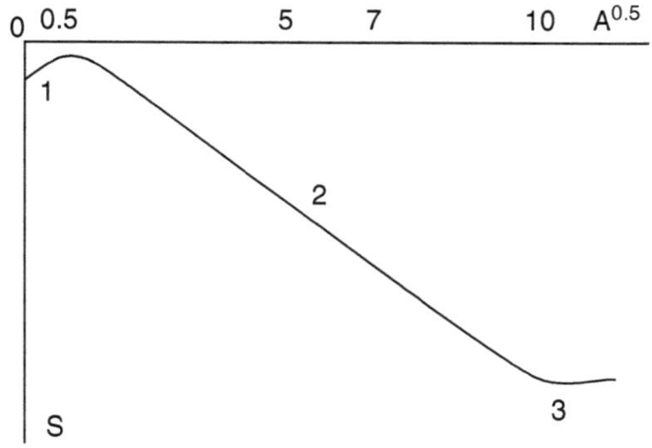

Fig. 1.6. Dependence of settling on footing area

is observed. We must also notice that the relation above is valid at pressures which do not exceed the soils practical limit of proportionality and at its enough homogeneity on a considerable depth.

At the same area of the footing, pressure and other equal conditions the settling of compact (circular, square etc.) foundations is higher than stretched ones. That follows also from analytical solutions (see further). A transfer from square footing to rectangular one (at equal specific pressure) decreases the active depth of the soil massif.

Influence of Load on Footing

At successive increase of a load on a soil three stages of its mechanical state are observed – of condensation 1 (Fig. 1.7), shearing displacement 2 and fracture 3. In the first of them the earth's volume decreases and deformation's rate falls with a tendency to zero. In this stage the dependence between the acting force F and the settling may be described by Hooke's law:

$$S = Fh/EA. \tag{1.8}$$

where E – modulus of elasticity.

The second stage is characterized by an appearance of shearing displacement zones with growth of which the settlings become higher and their rate decreases more slowly. In the third stage strains increase rapidly and soil expanses out of a footing. Deformation grows catastrophically and the settlings are big.

At cycling loading the soil's deformation increases with the number of cycles. Its elastic part changes negligibly and the full settling tends to a constant value. In the last state a soil becomes almost elastic.

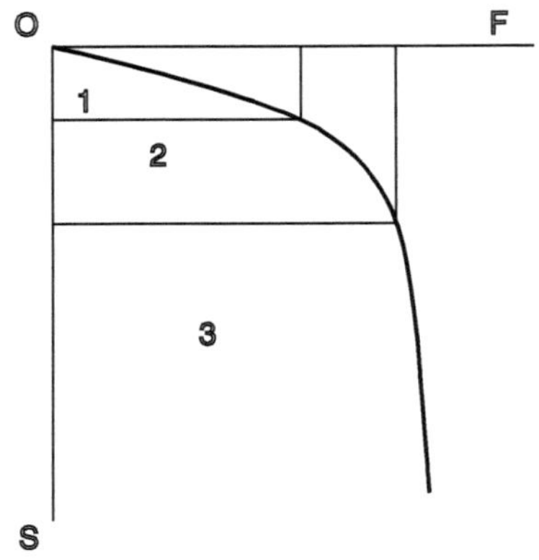

Fig. 1.7. Influence of load on footing

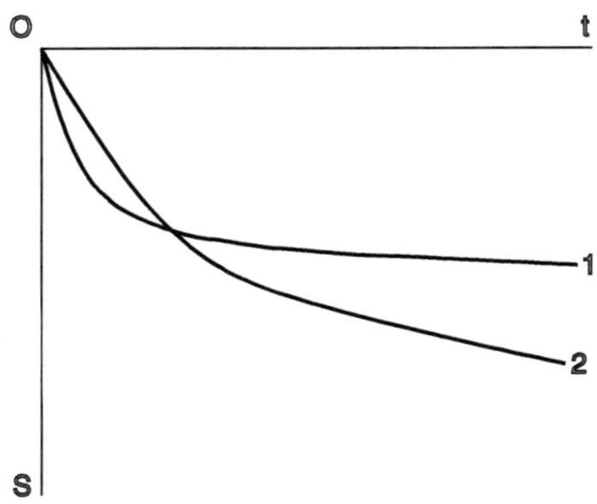

Fig. 1.8. Settling in time

Influence of Time

The experiments with earths and natural observations show that at constant load a development of the settling in time may be represented by Fig. 1.8. Curve 1 corresponds to sands in which settling happens fast as the resistance to squeezing a water out is small. Case 2 takes place in disperse soils such as clays, silts and others whose pores in natural conditions are filled with a water. The rate of the soil's stabilization depends on its water penetration and a creep of the skeleton.

The settling does not end in a period of a structure's construction and continues after it. The time at which the full settling takes place depends on the consolidation of a layer under the footing. In its turn the last phenomenon is determined by a rate and a character of external loading and by properties of soil, firstly by its compressibility and ability to water penetration. In conditions of good filtration the settling goes fast but at a weak water penetration the process may continue years.

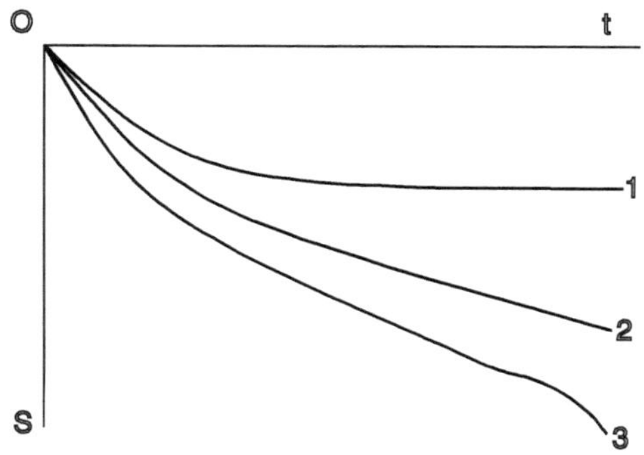

Fig. 1.9. Combined influence of time and loading

Combined Influence of Time and Loading

At small loads F the settling grows slowly (curve 1 in Fig. 1.9) and tends to a constant value. There is a maximum load on footing at which this process takes place. At bigger F the settling increases faster (line 2 in the figure) at approximately constant velocity and it may lead to a failure of structures (curve 3 in Fig. 1.9).

The velocity of the settling influences the strength of structures as they have different ability to redistribute the internal forces at non-homogeneous settlings of a footing. At high velocities of the settling the brittle fractures may take place, at slow ones – creep strains. For soils whose pores are fully filled with water the theory of filtration consolidation is usually used.

1.4.3 Computation of Settling Changing in Time

Premises of Filtration Consolidation Theory

The initial hypothesis of the theory is an assessment that the velocity of the settling's decrease depends on an ability of the water to penetrate the soil. Above that the following suppositions are introduced:

a) The pores of a soil are fully filled with water, which is incompressible, hydraulically continuous and free,
b) Earth's skeleton is linearly deformable and fully elastic,
c) The soil has no structure and initially an external pressure acts only on the water,
d) A water filtration in the pores subdues to the law of Darcy,
e) The compression of the skeleton and a transfer of the water are vertical.

Model of Terzaghi-Gersewanov

The model is a vessel (Fig. 1.10) filled with water and is closed by a sucker with holes. It is supported by a spring which imitates a skeleton of the soil

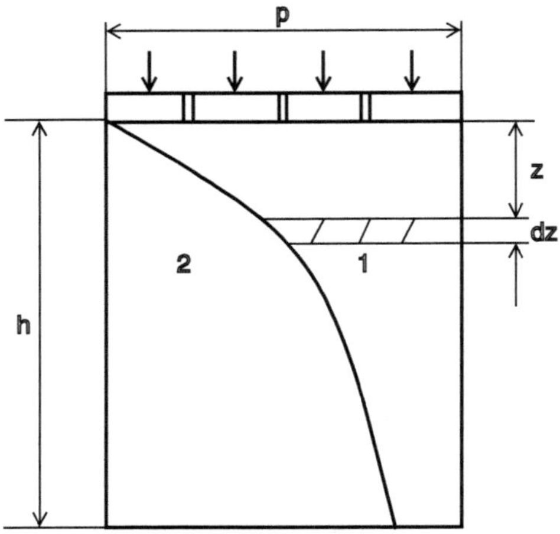

Fig. 1.10. Terzaghi-Gersewanov's model

and the holes – capillaries in earth. If we apply to the sucker an external load then its action presses in an initial moment only the water. After some time when a part of the water flows out of the vessel the spring begins to resist to a part of the pressure. The water is squeezed out slowly and has an internal pressure

$$H = p/\gamma_w \tag{1.9}$$

where γ_w is a specific gravity of the water.

The filtration of the water subdues to Darcy's law, which is however complicated by a presence of a connected water. At small values of the hydraulic gradient a filtration can not overcome the resistance of the water in the pores. Its movement is possible only at an initial value of the gradient.

Differential Equation of Consolidation Due to Filtration

In the basis of the theory the suppositions are put that a change of the water expenditure subdues to the law of filtration and a change of porosity is proportional to the change of the pressure.

Now we consider a process of the soil compression under a homogeneously distributed load. We suppose that in an initial state the soil's massif is in a static state which means that a pressure in pores is equal to zero. We denote the last as p_w (zone 2 in the figure) and effective pressure which acts on the solid particles as p_z (zone 1 in the figure). At any moment the sum of these pressures is equal to the external one as:

$$p = p_w + p_z. \tag{1.10}$$

In time the pressure in water decreases and in the skeleton – increases until the last one supports the whole load.

At any moment an increase of the water expenditure q in an elementary layer dz is equal to the decrease of porosity n that is

$$\partial q/\partial z = -\partial n/\partial t. \tag{1.11}$$

The last expression gives the basis for the inference of differential equation of a consolidation theory which with consideration of (1.9) is

$$\partial H/\partial t = C_v \partial^2 H/\partial z^2 \tag{1.12}$$

where $C_v = K_\infty a_v \gamma_w$ is a factor of the soil's consolidation.

Relation (1.12) is a well-known law of diffusion and it is usually solved in Fourier series.

For the determination of a settling in a given time a notion of consolidation degree is usually used. It is defined as the ratio between the settling in the considered moment and a full one or

$$u = S_f/S. \tag{1.13}$$

It may be found through the ratio of areas of pressure diagrams in the skeleton (p_z) at the present moment and at infinite time as

$$u = \int_0^h (p_z/A_p)dz \tag{1.14}$$

where A_p is the area of fully stabilized diagram of condensed pressure.

Putting in (1.13) expression for pressure p_z in the soil's skeleton which is received at the solution of equation (1.12) we have after integration

$$u = 1 - 8(e^{-N} + e^{-9N}/9 + e^{-25N}/25 + \ldots..)/\pi^2 \tag{1.15}$$

where e is the Neper's number, $N = \pi^2 C_v t/4h^2$ – a dimensionless factor of time. Putting (1.15) in (1.7) we find the settling at given time t.

For bounded, hard plastic and especially of firm consolidated soils, containing connected water the theory cannot be used.

In the conclusion we must say that (1.15) is the creep equation and it is difficult to apply it to boundary problems and for this task simpler rheological laws can be used and this way is considered in the book.

1.5 Description of Properties of Soils and Other Materials by Methods of Mechanics

1.5.1 General Considerations

Usually problems of the Soil Mechanics are divided in two main parts.

The first of them deals with the settling of structures due to their own weight and other external forces. This problem is almost always solved by the methods of the Elasticity Theory although many earths show non-linear residual strains from the beginning of a loading.

The second task is the stability which is connected with an equilibrium of an ideal soil immediately before an ultimate failure by plastic flow. The most important problems of this category are the computation of the maximum pressure exerted by a massif of soil on elastic supports, the calculation of the ultimate resistance of a soil against external forces such as a vertical pressure acting on an earth by a loaded footing etc. The conditions of a loss of the stability can be fulfilled only if a movement of a structure takes place but the moment of its beginning is difficult to predict.

So, we must consider the conditions of loading and of support required to establish the process of transition from the initial state to a failure. Here we demonstrate on simple examples the approaches that can help to study this process.

1.5.2 The Use of the Elasticity Theory

Main Ideas

Some earths components and even their massifs (rock, compressed clays, frozen soils etc) subdue to the Hooke's law i.e. for them displacements are proportional to external forces almost up to their fracture without residual strains. Here in a simple tension (Fig. 1.11) stress p which is determined by relation (1.1) and is equal in this case to normal component σ is linked with relative elongation

$$\varepsilon = l/l_o - 1 \tag{1.16}$$

where index o refers to initial values (broken lines in the figure) by the law similar (1.8):

$$\sigma = E\varepsilon \tag{1.17}$$

Here according to the main idea of the book we show the use of this expression for the prediction of a failure of some structure elements.

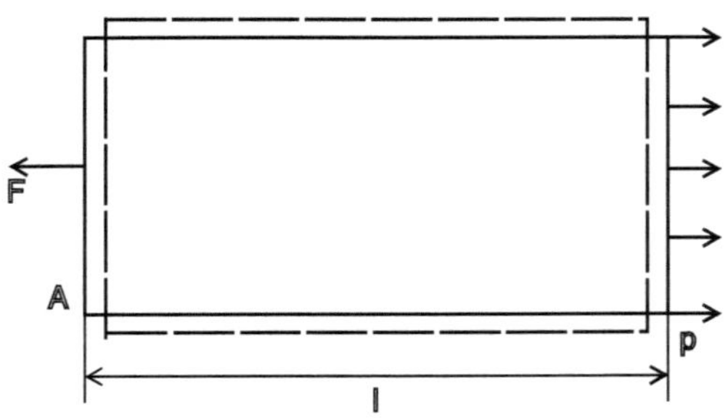

Fig. 1.11. Bar in tension

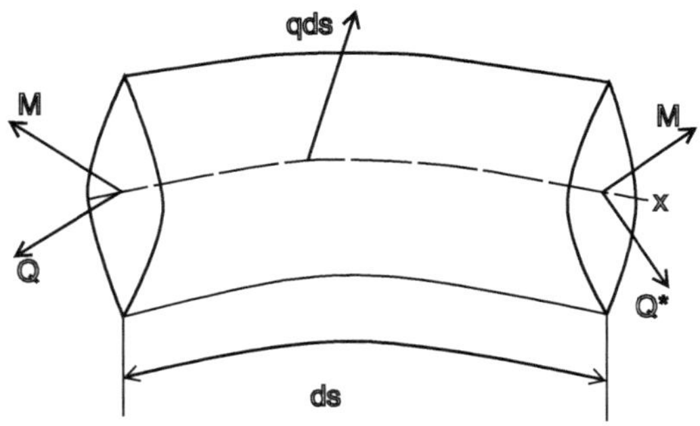

Fig. 1.12. Element of bar under internal and external forces

Some Solutions Connected with Stability of Bars

As was told before one of the first methods of theoretical prediction of failure gave L. Euler for a compressed bar. His approach can be generalized and we give the final results. We begin with static equations of a part of it (Fig. 1.12) as

$$d\delta Q/ds + \chi_o x\delta Q + \delta\chi x Q_o = 0, d\delta M/ds + \chi_o x\delta M + \delta\chi x M_o + ix\delta Q = 0 \quad (1.18)$$

and constitutive equation similar (1.17):

$$\partial M = B_j E\delta\chi (j = x, y, z). \quad (1.19)$$

Here i, Q, M are vectors – unit, of shearing force, of bending moment, $Q^* = Q + dQ$, $M^* = M + dM$, δ – sign of their increment, χ – curvature of the bar, x – sign of multiplication, $B_j E$ – rigidity and subscript $_o$ denote initial values (before a failure).

For a ring with radius r and thickness h under external pressure q we have $M_o = \chi_{xo} = \chi_{yo} = Q_{yo} = Q_{zo} = 0$, $\chi_{yo} = 1/r$, $Q_{xo} = -qr$, $ds = rd\theta$ where z, x are normal and tangential directions, θ is the second polar co-ordinate. For the planar loss of stability we have from (1.18), (1.19)

$$d^2\delta Q_z/d\theta^2 + (1 + qr^3/B_y E)\delta Q_z = 0.$$

The solution of this problem is well-known and the maximum critical pressure is

$$q_* = 3EI_y/r^3.$$

This result is valid for a long tube if we replace moment of inertia I_y relatively to axis y by cylindrical rigidity $h^3/12(1 - \nu^2)$ where ν is the Poisson's ratio. The solution is used for the appreciation of the strength of galleries in an earth.

For a bar under compression and torsion (Fig. 1.13) the solution of (1.18), (1.19) can be presented in the form

$$(M/M_*)^2 + F/F_* = 1 \quad (1.20)$$

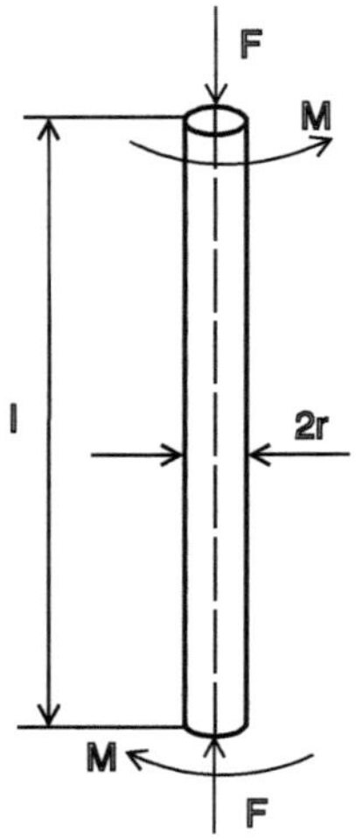

Fig. 1.13. Circular bar under compression and torsion

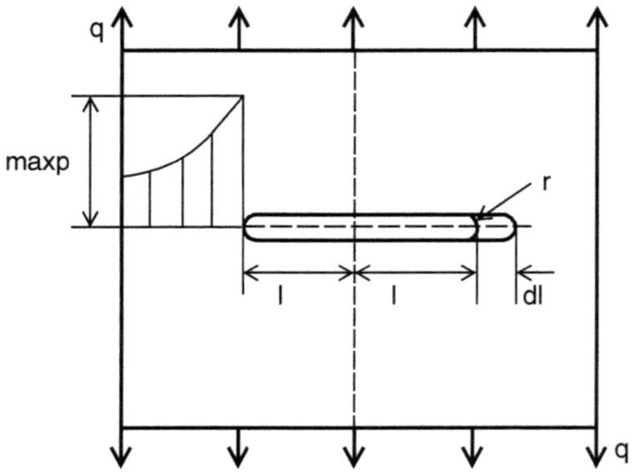

Fig. 1.14. Crack under tension

where values $M_* = 2\pi EI/l$, $F_* = \pi^2 EI/l^2$ refer to the separate actions of the respective loads, $I = \pi r^4/4$ is axial moment of inertia of a circle. The relation can be used for the appreciation of strength of bores, piles, drills etc.

Bases of Crack Mechanics

The basis of the modern linear mechanics of cracks was given by English scientist A. Griffith /8/. In order to clear up the idea we consider a plate (Fig. 1.14) with a narrow crack of length 2l perpendicular to tensile stresses q in infinity. The distribution of stresses near the crack's edge is shown in the left part of the picture and their maximum can be given by

$$\max p = 2q\sqrt{l/r} \tag{1.21}$$

where r is a radius of curvature in the end of the crack. When the last one begins to propagate the following amount of the work is freed:

$$dW = (\max p)^2 (r/E) dl$$

or with consideration of (1.21)

$$dW = 4(q^2/E) l dl. \tag{1.22}$$

This process is resisted by the forces of surface stretching with an energy $dU = 2\gamma_s dl$ where γ_s is the energy per unit length. In the critical state $dU = dW$ and we derive from (1.22) the value of a critical stress:

$$q_* = \sqrt{\gamma_s E/2l}. \tag{1.23}$$

The more detailed analysis of this theory will be given in Chaps. 3–5.

1.5.3 The Bases of Ultimate Plastic State Theory

Main Ideas

Many soils and other materials have small hardening and angle of internal friction as well as negligible deviation from the condition of constant volume. For them the ultimate state of structures according to the scheme of an ideal (perfect) elastic-plastic body can be used. Its diagram in coordinates p, e is given in Fig. 1.15 with p^* – an yielding point. In the Media Mechanics consequent coordinates are σ, ε and σ_{yi}.

To catch the idea of the method we consider a part of a beam (Fig. 1.16, a) loaded by moments M with a cross-section on Fig. 1.16, b. Diagram of stress in elastic state with yielding value at most remote point of the compressed part of the cross-section is given in Fig. 1.16, c. With the growth of the moment σ-diagram in the compressed zone becomes a trapezoid. Then the yielding point is reached in the tensile zone and the σ-diagram there becomes the trapezoid too.

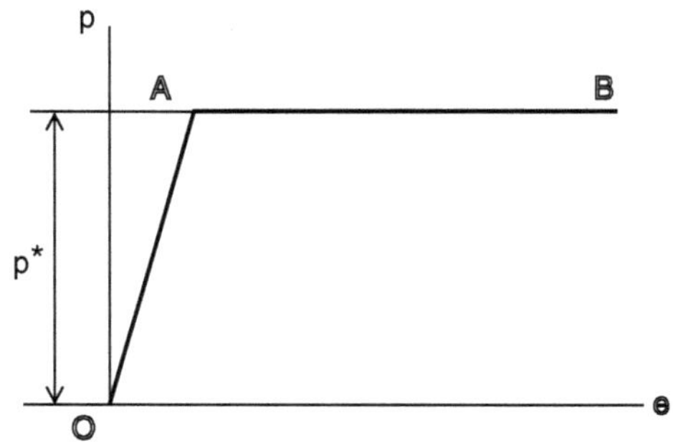

Fig. 1.15. Diagram of ideal elastic-plastic material

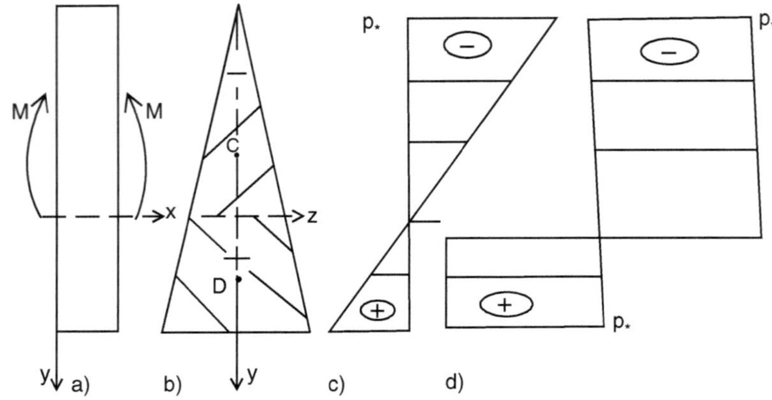

Fig. 1.16. Pure bending of beam

The distribution of stresses at ultimate state is drawn in Fig. 1.16, d where we have two rectangles. From equilibrium condition we receive

$$\Sigma X = \sigma_{yi} A_+ - \sigma_{yi} A_- = 0$$

from which we have $A_+ = A_- = A/2$. It means that in ultimate state tensile and compressed areas are equal to one half of the whole one. From the second static equation ($\Sigma M_z = 0$) we find the ultimate load as

$$M_* = \sigma_{yi} A(CD)/2$$

where (CD) is the distance between centroids of compressed and tensile zones. Particularly for a rectangle with a width b and a height h we compute

$$M_* = \sigma_{yi} bh^2/4. \tag{1.24}$$

Here we must underline that we used for the final results only the horizontal part of the diagram in Fig. 1.15 and so the method may be applied to a brittle fracture too.

Ultimate State of Statically Indetermined Beams

As the second practical example of the theory we consider a retaining wall or a piles sheet under triangular load of a soil or a liquid (Fig. 1.17, a). The approximate M-diagram in elastic state is given in Fig. 1.17, b. In the ultimate state plastic hinges with moments M_* (see relation (1.24) at $b = 1$) appear. According to M-diagram one of them (see the broken line in Fig. 1.17, a) is in the fixed end (point A), another – somewhere in the span (point C). In order to find distance x we use static equations

$$-M_A = ql^2/6 - Hl = M_*, \quad M_C = Hx - qx^3/6l = M_*.$$

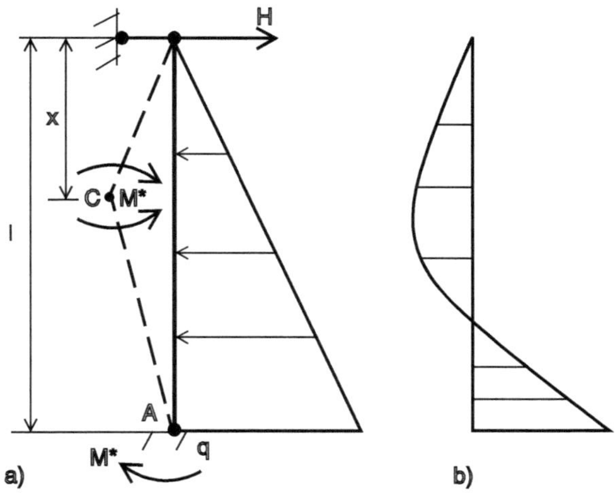

Fig. 1.17. Statically indetermined beam

Excluding from these two expressions reaction H we have

$$q = 6M_*/x(l - x).$$

According to the second Gvozdev's theorem /9/ the position of hinge C gives minimum to q-value. So, we compute

$$x = l/2, q_* = 24M_*/l^2.$$

The same value of q_* can be got according to the first Gvozdev's theorem as a maximum load in the ultimate static state /7/.

Ultimate State of Plates in Bending

We consider firstly a polygonal plate (Fig. 1.18, a) with simply supported edges. We model its part ODCE as a double-supported beam AB (Fig. 1.18, b) with a M-diagram similar to the broken axis (broken line in Fig. 1.18, c) under concentrated load in point O which acts on the plate. The value of M_{max} is evident

$$M_{max} = M_C = Fd(l - d)/l. \tag{1.25}$$

Taking M_{max} equal to M_* according to formula (1.24) at b = 1, substituting in (1.25) $d = L\tan\varphi_s$, $l - d = L\tan\psi_s$ where $\varphi_s = a$, $\psi_s = c$ in Fig. 1.18, a, $s = 1, 2, \ldots n$ (n is a number of plate's corners), and the moments along the rays of fracture from point O to the corners of the plate we find

$$F_* = M_* \sum_{s=1}^{n} (\cot\varphi_s + \cot\psi_s). \tag{1.26}$$

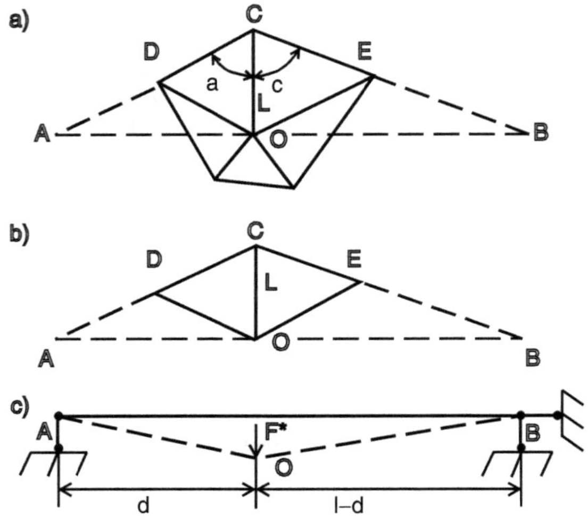

Fig. 1.18. Polygonal plate in bending

For a rectangle with a width b, a height h we have $\cot \varphi_s = h/b$, $\cot \psi_s = b/h$, $n = 4$ and

$$F_* = 4M_*(h/b + b/h).$$

In the case of a square ($h = b$) it gives

$$F_* = 8M_*. \tag{1.27}$$

For a right polygon with n corners and the force F in its centre we have $\varphi_s = \psi_s = \pi(0.5 - n^{-1})$, $\cot \varphi_s = \cot \psi_s = \tan(\pi/n)$ and according to (1.26)

$$F_* = 2nM_* \tan(\pi/n).$$

When $n \to \infty$ we derive for a circle

$$F_* = 2\pi M_*. \tag{1.28}$$

If the plate has fixed edges we can suppose that the fracture occurs along them too. From Fig. 1.19 where part ODE of the plate is represented we have the static equation $\Sigma M_{DE} = 0$

$$F_* H = M_* l$$

and as $d = H \cot \varphi_s$, $l - d = H \cot \psi_s$ we get again (1.26) and so for our case

$$F_* = 2M_* \sum_{s=1}^{n} (\cot \varphi_s + \cot \psi_s). \tag{1.29}$$

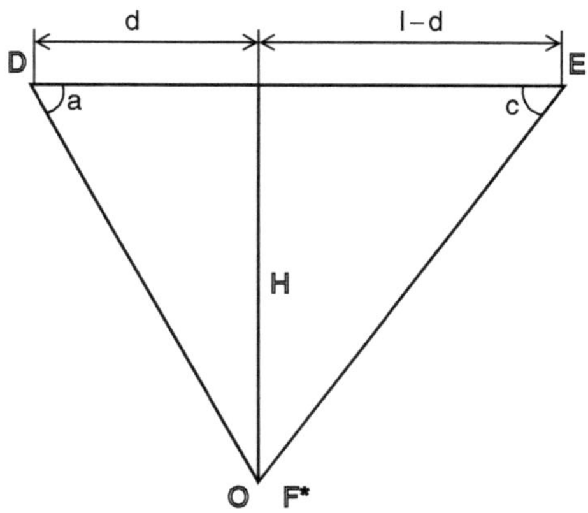

Fig. 1.19. Equilibrium of part of plate

For a circle with the critical force in its centre we have similar to (1.28)

$$F_* = 4\pi M_* \tag{1.30}$$

and it is less than in the case of a right polygon $F_* = 4nM_* \tan(\pi/n)$ and so its mode of fracture is a circle of an indeterminate radius.

If the distributed load q is applied to the plate we can suppose that it does the same work as its resultant F. So, in the case of q = constant we must replace in the previous relation force F_* by $q_*A/3$ where A is an area of the plate's surface. So, for a right polygon with supported and fixed ends as well as for a circle we have respectively

$$q_* = 6M_*/R^2, q_* = 12M_*/R^2 \tag{1.31}$$

where R is radius of an internal circumference. The first (1.31) is valid for a square $2R \times 2R$.

Another use of the first Gvozdev's theorem for plates will be given later in Chap. 3.

1.5.4 Simplest Theories of Retaining Walls

We consider a development of main ideas of the ultimate state computation on the example of a retaining wall loaded mainly by a soil. The first investigators of this problem supposed that in an ultimate state triangle ABC (Fig. 1.20, a) is against the wall. After somewhat naïve work of Belidor (1729) Coulomb /11/ considered the earth pressure on the portion BC of the vertical side CE and assumed that the earth has a tendency to slide down along some plane AB. Neglecting any friction on CB he concluded that reactions of the wall with their resultant H are horizontal. Weight Q of prism ABC is

$$Q = (\gamma_e h^2 \tan \Psi)/2$$

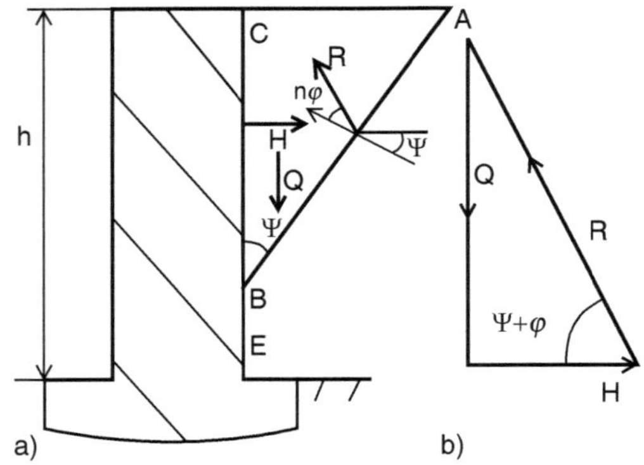

Fig. 1.20. Retaining wall

where γ_e is a specific weight of the earth. Resultant R of the reactions along sliding plane AB forms friction angle φ with normal n to AB. It means that R is inclined to a horizontal direction under angle $\Psi + \varphi$.

The triangle in Fig. 1.20, b represents the condition of an equilibrium of prism ABC from which we have

$$H = Q\cot(\Psi+\varphi) = 0.5\gamma_e h^2 \tan\Psi \cot(\varphi+\Psi) = 0.5\gamma_e(1-f\tan\Psi)/(1+f\cot\Psi) \quad (1.32)$$

where $f = \tan\varphi$ is a coefficient of friction. Then he found the maximum of H and equalling its derivative by Ψ to zero he received expression

$$\tan^2\Psi = (1 - f\tan\Psi)/(1 + f\cot\Psi)$$

from which it follows

$$\tan\Psi = -f + \sqrt{1+f^2}. \quad (1.33)$$

Prony threw equation (1.33) in simpler form, viz $\cot(\Psi+\varphi) = \tan\Psi$ which gives

$$\Psi = 0.5(\pi/2 - \varphi). \quad (1.34)$$

All above was concerned an active pressure of a soil on the retaining wall. Similar computations for a passive state of an earth when the wall moves against it gives relation like (1.34)

$$\Psi = 0.5(\pi/2 + \varphi). \quad (1.35)$$

Rankine /12/ offered a method of finding proper dimensions of a retaining wall. He considered a horizontal plane (Fig. 1.21) when σ_x, σ_y are main stresses. It allows to construct Mohr's circle (Fig. 1.22) in coordinates τ, σ. In ultimate state φ is an angle of repose, so from the figure we have

$$\sigma_y - \sigma_x = (\sigma_y + \sigma_x)\sin\varphi$$

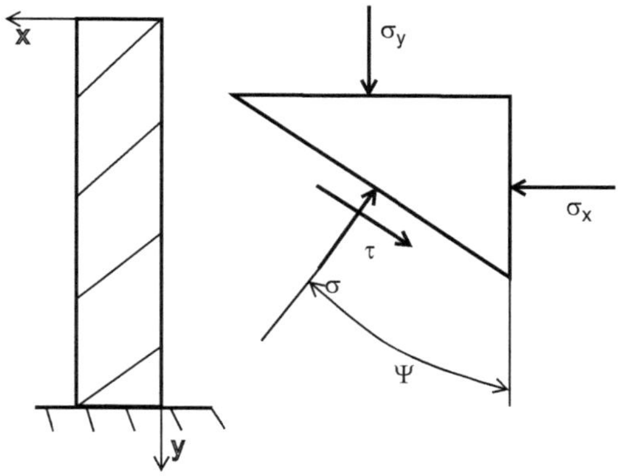

Fig. 1.21. Stress state near retaining wall

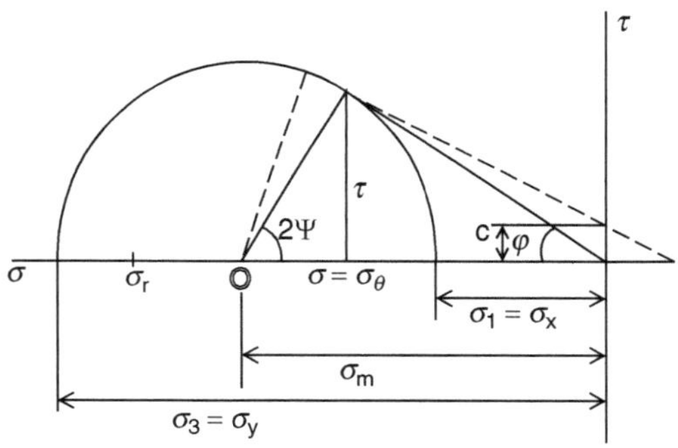

Fig. 1.22. Mohr's circle

or

$$\sigma_y/\sigma_x = (1 + \sin\varphi)/(1 - \sin\varphi). \qquad (1.36)$$

But since $\sigma_y = \gamma_e y$ the maximum horizontal reaction on the retaining wall in ultimate equilibrium is

$$\sigma_x = \gamma_e y(1 - \sin\varphi)/(1 + \sin\varphi). \qquad (1.37)$$

Rankine recommends this pressure as to be used when investigating a stability of a retaining wall. In the case of a passive pressure similar computations give

$$\sigma_x = \gamma_e y(1 + \sin\varphi)/(1 - \sin\varphi). \qquad (1.38)$$

Relations (1.37), (1.38) are very often used in design practice (see Chap. 4). But it is not difficult to show that they coincide with the Coulomb's law (1.32) if we put there the value of Ψ according to (1.34), (1.35). So, we can conclude that the Rankine's angle of repose is equal to the Coulomb's angle of friction and their theories are the same.

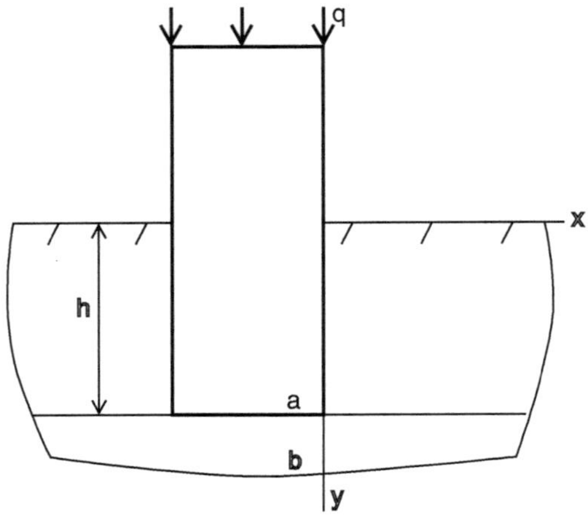

Fig. 1.23. Depth of retaining wall

A theory for determining the necessary depth of foundation was proposed by Pauker (St.-Petersburg, 1889) – Fig. 1.23. Considering an element ab under the wall and using equation (1.37) of the Rankine's theory he concluded that at the moment of sliding the lateral pressure σ_x must satisfy equation

$$\sigma_x = q(1 - \sin\varphi)/(1 + \sin\varphi). \tag{1.39}$$

Taking into account relation (1.38) for an ultimate state of the soil with depth h instead of y and combining it with (1.39) he get the necessary depth of the foundation as

$$h = (q/\gamma_e)(1 - \sin\varphi)^2/(1 + \sin\varphi)^2. \tag{1.40}$$

To verify this equation some interesting experiments were provided by V. Kourdumov in the laboratory of the Ways Communication institute in St.-Petersburg and by H. Müller-Breslau in 1906. These tests showed that the pressure on the wall can sometimes be higher than that predicted by the Coulomb's theory.

1.5.5 Longtime Strength

All the materials deform in time and this phenomenon is called a creep. The creep curves for soils are shown in Figs. 1.8, 1.9 and in the Media Mechanics they are considered in coordinates ε, t at different σ. We can conditionally distinguish three parts of creep- a primary one with $d^2\varepsilon/dt^2 < 0$, the second where $d\varepsilon/dt$ is approximately constant and the third with $d^2\varepsilon/dt^2 > 0$. The last one is usually linked with fracture due to a damage (a development of internal defects) and a decrease of cross-section (at tension).

N. Hoff /13/ used for the second portion of the curves a power law

$$d\varepsilon/dt = B\sigma^m, \tag{1.41}$$

where B-coefficient and m is an exponent of the hardening law, to predict the rupture time at tension which corresponds to infinite elongation of a bar under constant load F. He used also a link between conditional $\sigma_o = F/A_o$ and true $\sigma = F/A$ stresses

$$\sigma = \sigma_o e^{\varepsilon}. \tag{1.42}$$

which follows from the condition of constant material's volume $A_o l_o = Al$ (Fig. 1.11). Here

$$\varepsilon = \ln(l/l_o). \tag{1.43}$$

is a true strain.

Putting (1.42) into (1.41) he received after integration in limits $0 \le t \le t_\infty$, $0 \le \varepsilon \le \infty$ a finite value of rupture time as

$$t_\infty = (B(\sigma_o)^m m)^{-1}. \tag{1.44}$$

According to this relation a structure can be destroyed at any load. It makes conditional the traditionally used strength limits. The scheme can be generalized.

Experiments show that for creep curves the following law may be introduced /7, 10/

$$\varepsilon e^{-\alpha\varepsilon} = \Omega(t)\sigma^m \tag{1.45}$$

where Ω is an experimentally determined function of time and factor $e^{-\alpha\varepsilon}$ ($\alpha \ge 0$) takes into account a damage increase that induces a decrease of ultimate strain and time for not enough ductile materials, a growth of their volume, the third parts of creep curves and other effects.

Tests show that relation (1.45) is valid for monotonous loading while strain ε does not diminish. At stepwise or interrupted change of stress (Fig. 1.24) rheological law (1.45) may be applied to the parts where σ is not less than on previous ones if the time is calculated from a beginning of a new application of stress. The agreement of (1.45) with an experiment is better for more unsteady creep /14/. It may be used for finding the ultimate state of structures.

Putting (1.42) into (1.45) and using the criterion /15/ $d\varepsilon/dt \to \infty$ we get on critical strain and time as

$$\varepsilon_* = (\alpha + m)^{-1}, \Omega(t_*) = ((\alpha + m)(\sigma_o)^m e)^{-1}. \tag{1.46}$$

If a change of the bar's dimension is small or true stress σ is constant we have from (1.46) the fracture values due to damage only as

$$\varepsilon_* = 1/\alpha, \Omega(t_*) = (\alpha\sigma^m e)^{-1}. \tag{1.47}$$

Similarly from (1.46) we can find at $\alpha = 0$ critical values for perfectly ductile bodies. When an influence of time is negligible (Ω = constant) law (1.45) becomes a constitutive equation of the Plasticity Theory generalized here and expressions (1.46), (1.47) develop the theory of maximum loads at unstable deformation.

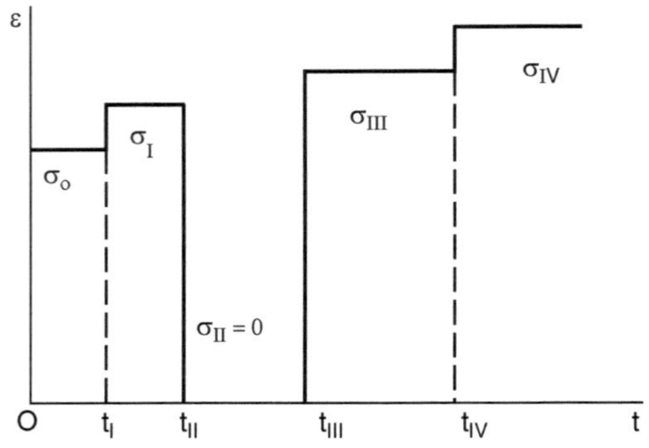

Fig. 1.24. Stepwise and interrupted loading

Expressions (1.45)–(1.47) give an opportunity to find all the parameters in rheological law (1.45). When we take from the primary parts of creep curves at the same time strains ε we can construct diagrams σ-ε whose approximation by the power function allows to find the exponent m of the hardening law. Dividing ε by σ^m we have a creep curve $\Omega(t)$ which can be also approximated by a power law e.g. $\Omega = Bt^n$ Damage factor. α may be found in two ways, either according to the first expressions (1.46), (1.47) or by approximation of the third parts of creep curves. Very often the both approaches give near values of α.

1.5.6 Eccentric Compression and Determination of Creep Parameters from Bending Tests

We take a bar with a width 2h (Fig. 1.25, a) and unity thickness loaded by compressive force F with eccentricity e. We use the law of plane cross-sections $\varepsilon = ky_1$ (broken line in the figure) where k is an angle of their rotation and two systems of coordinates – xOy and xO_1y_1 with $y_1 = y + c$. If we apply forces in point O we find axial load N = F and bending moment M = Fe.

We take a material of non-linear creep type (1.45) at $\alpha = 0$ as

$$\sigma = \omega(t)\varepsilon^{\mu} = p \tag{1.48}$$

where $\omega = \Omega^{-\mu}$, $\mu = 1/m$ $(0 \leq \mu \leq 1)$. At $\omega =$ constant and $\mu = 1$ (then $\omega = E$) we have the Hooke's law (1.17) – straight line OA in Fig. 1.15. If $\mu = 0$ we receive the perfect plastic body with $\omega = \sigma_{yi} = p^*$ – horizontal straight line in the same figure.

With consideration of the expression above for ε we derive from (1.48) – Fig. 1.25, b.

$$p = \omega k^{\mu}(y_1)^{\mu} = \omega k^{\mu}(y + c)^{\mu} \tag{1.49}$$

Putting (1.49) into static equations $\Sigma X = 0$, $\Sigma M_{o1} = 0$ we compute

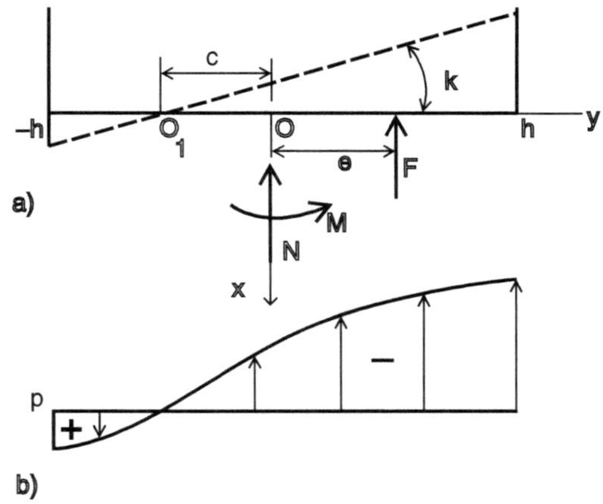

Fig. 1.25. Eccentric compression of bar

$$N = \omega k^{\mu} \int\limits_{h-c}^{h+c} (y_1)^{\mu} dy_1 = \omega k^{\mu}((h+c)^{1+\mu} - (h-c)^{1+\mu})/(1+\mu), \quad (1.50)$$

$$M = \omega k^{\mu} \int\limits_{h-c}^{h+c} (y_1)^{1+\mu} dy_1 - Nc$$
$$= \omega k^{\mu}((h+c)^{2+\mu} + (h-c)^{2+\mu})/(2+\mu) - Nc. \quad (1.51)$$

Now we consider some particular cases for ideal materials that are well-known from a literature on the Strength of Materials.

If $\mu = 1$, $\omega = E$ we have from (1.50), (1.51) $N = 2Ekch$, $M = 2Ekh^3/3$ and from (1.49) – a well-known relation (generalization of this law for brittle material see in Appendix A)

$$p = N/2h + 3My/2h^3.$$

When p at $y = h$ reaches its yielding point p^* we receive the condition (straight line in Fig. 1.26 in coordinates M/p^*h^2, $N/2p^*h$)

$$N/2p^*h + 3M/2p^*h^2 = 1. \quad (1.52)$$

At $\mu = 0$, $\omega = p^*$ we receive from (1.50), (1.51) $N = 2p^*c$, $M = p^*(h^2 - c^2)$ and after the exclusion of c we have (solid curve in the figure)

$$M/p^*h^2 + (N/2p^*h)^2 = 1. \quad (1.53)$$

It is interesting to notice a similarity of (1.53) and (1.20) that shows a resemblance of the behaviour of structures in an ultimate state.

Such simple results can not be received for other μ. However we can put different c in relations (1.49)–(1.51) and for $p^* = p(h)$ after an exclusion

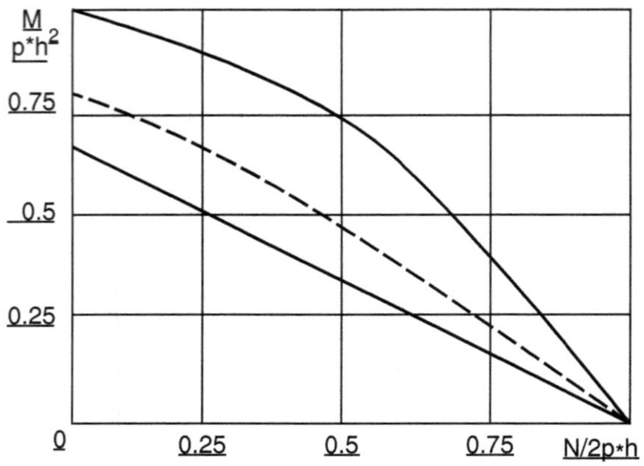

Fig. 1.26. Ultimate state of bar in eccentric compression-tension

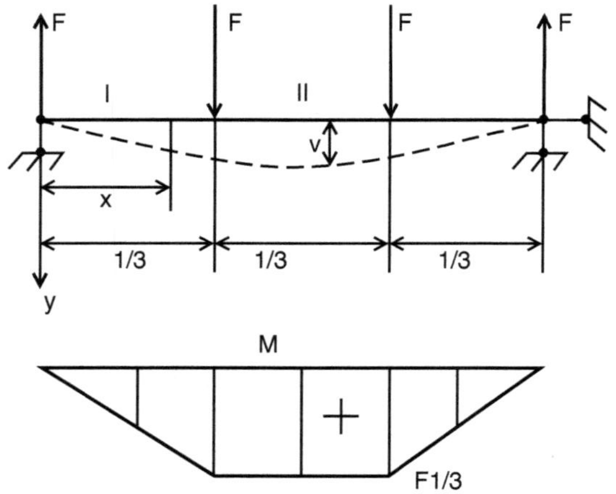

Fig. 1.27. Deflection of beam

of ωk^{μ} we get $N/2p^{*}h$, $M/p^{*}h^{2}$. Obtained by this way curve for $\mu = 0.5$ is constructed in Fig. 1.26 by broken line.

If $c = 0$ we find from (1.50), (1.51) $N = 0$, $M = \omega k^{\mu} I$ which gives a generalization of approximate bending equation

$$v'' = -M^{m}/(\omega I)^{m} \tag{1.54}$$

where $I = 2h^{2+\mu}/(2+\mu)$ – moment of inertia relatively to axis z perpendicular to directions x, y in Fig. 1.27, v – deflection of a beam and sign ʻdenotes the derivative by coordinate x.

Function v(x) can be found by the same procedure as in the Strength of Materials and the Structural Mechanics for linear material. However, here we can not use a single equation for the whole beam and must consider its parts separately. We shall demonstrate it on the scheme which is often used in tests for a study of materials behaviour (Fig. 1.27). Since the beam is symmetric we can consider only its two portions. So, at $(F/\omega I)^m = K$ we have in parts I and II

$$\begin{aligned} v_I'' &= -Kx^m,\ v_I' = v_I'(0) - Kx^{m+1}/(m+1), \\ v_I &= v_I'(0)x - Kx^{m+2}/(m+1)(m+2), \\ v_{II}'' &= -K(l/3)^m,\ v_{II}' = v_{II}'(0) - Kx(l/3)^m, \\ v_{II} &= v_{II}(0) + v_{II}'(0)x - Kx^2(l/3)^m/2. \end{aligned} \tag{1.55}$$

Here we take $v_I(0) = 0$. From condition $v_{II}'(l/2) = 0$ we find

$$v_{II}'(0) = Kl^{m+1}/3^m 2.$$

From border demand $v_I'(l/3) = v_{II}'(l/3)$ we compute

$$v_I'(0) = K(l/3)^{m+1}(m+3)/2(m+1)$$

Now the similar condition $v_I(l/3) = v_{II}(l/3)$ gives the last unknown as

$$v_{II}(0) = -K(l/3)^{m+2}m/2(m+2).$$

So, all the constants in (1.55) are found and the displacement and the angle of rotation in any point of axis x are determined. Particularly, the deflection in the middle of the beam where it is usually measured is

$$v_{II}(l/2) = K(l/3)^{m+2}(5m+18)/8(m+2) \tag{1.56}$$

At $m = 1$ we compute well-known result $v(l/2) = 23Fl^3/648EI$. Approximation of (1.56) gives m, ω and hence $\Omega(t)$.

1.5.7 Fracture of Tunnel Arch

We begin with the task of a pure bending of a large curvature rectangular rod. We take the unity width of it and height 2h (Fig. 1.28,a). Let ρ_o, ρ are curvature radii of the bar axis and of the neutral layer. The strain of fibres at distance x from axis y is evident from Fig. 1.28b as

$$\varepsilon = x(\Delta d\varphi/d\varphi)/(x+\rho). \tag{1.57}$$

By condition (1.48) we receive expression for normal stress as follows

$$\sigma = \omega(t)(\Delta d\varphi/d\varphi)^\mu (x/(x+\rho))^\mu. \tag{1.58}$$

At $\mu = m = 1$ we have linear approach and S. Timoshenko /5/ showed that in the following range

$$1.3 < (\rho_o + h)/(\rho_o - h) < 3$$

it is in a good agreement with the Theory of Elasticity solution.

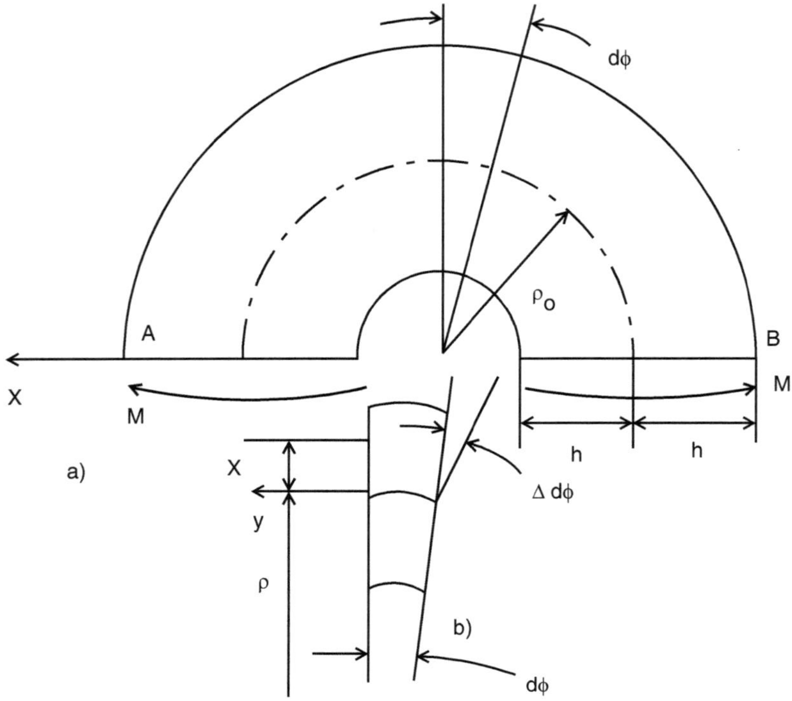

Fig. 1.28. Bending of large curvature rod

Developing the linear approach we determine the position of the neutral axis from condition $\Sigma Y = 0$ that with consideration of (1.58) gives

$$\int_{e-h}^{e+h} (x/(x+\rho))^{\mu} dx = 0 \tag{1.59}$$

where $e = \rho_o - \rho$. In the common case ρ must be found by attempts method beginning, e.g., from its value at $\mu = 1$. For $\mu = 0.5$ integral in (1.59) can be computed rigorously giving condition

$$\sqrt{(e+h)(\rho_o+h)} - \sqrt{(h-e)(\rho_o-h)}$$
$$-\rho \ln\left(\left(\sqrt{\rho_o+h}+\sqrt{e+h}\right)/\left(\sqrt{\rho_o-h}+\sqrt{h-e}\right)\right) = 0. \tag{1.60}$$

Calculations at $\rho_o = 3\,h$ give $\rho = 2.92\,h$ which is near to consequent value by the linear solution /47/:

$$\rho = 2h/\ln((\rho_o+h)/(\rho_o-h)) = 2.8854h.$$

To find the stresses we write condition $\Sigma M_z = 0$ which with consideration of (1.58) gives

$$M = \omega(t)J(\Delta d\varphi/d\varphi)^{\mu}. \qquad (1.61)$$

Here

$$J = \int_{e-h}^{e+h} (x^{1+\mu}/(x+\rho)^{\mu})dx$$

Excluding from (1.58), (1.61) multiplier $(\Delta d\varphi/d\varphi)^{\mu}$ we derive

$$\sigma = Mx^{\mu}/J(x+\rho)^{\mu}. \qquad (1.62)$$

In the common case integral J must be calculated approximately, however at $\mu = 0.5$ it can be represented with consideration of (1.60) in form

$$J = 0.5\left((e+h)^{3/2}\sqrt{\rho_o + h} - (h-e)^{3/2}\sqrt{\rho_o - h}\right).$$

The stresses in points A, B at Fig. 1.29 are

$$/\sigma_A/ = M\sqrt{(e+h)(\rho_o + h)}/J = 1.0422M/h^2,$$
$$/\sigma_B/ = M\sqrt{(h-e)(\rho_o - h)}/J = 1.3603M/h^2. \qquad (1.63)$$

Diagrams of stress distribution according to (1.5.7) are given in Fig. 1.29 by solid line.

It is interesting to compare the results received to the solution for straight bar. Putting the stresses like (1.49)

$$\sigma = \omega(t)k^{\mu}x^{\mu}$$

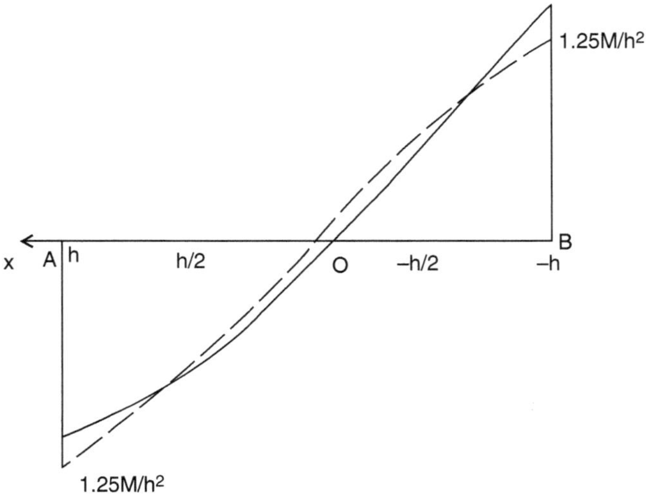

Fig. 1.29. Diagrams of stresses

into condition $\Sigma M_z = 0$ we have after exclusion of ωk^μ

$$\sigma = M(2+\mu)x^\mu/2h^{2+\mu}. \tag{1.64}$$

From Fig. 1.29 where by broken line also for $\mu = 0.5$ the curve according (1.64) and data from /5/ are represented we can see that with a fall μ the influence of the curvature on bearing capacity of the bar is reducing.

The tensile stresses are the most dangerous and at certain direction of the bending moment they can have the absolute highest value. For the case of small strains expressions (1.46) give

$$\varepsilon_* = 1/\alpha,\ \Omega(t_*) = (\alpha e \sigma_+^m)^{-1},$$

Here σ_+-the maximum tensile stress. S. Timoshenko in /5/ recommends the linear option of the solution for computation of tunnel arches strength and, so, the results above have the meaning for the Geomechanics.

2

Main Equations in Media Mechanics

2.1 Stresses in Body

The stresses p, σ, τ for tension and compression were introduced in Chap. 1. In order to study the stress state in a point the cube is usually cut about it (Fig. 2.1). Acting on its faces stresses are often written in a form of a matrix which is called the tensor (of stresses) as

$$T_\sigma = \begin{pmatrix} \sigma_{xx} & \tau_{xy} & \tau_{xz} \\ \tau_{yx} & \sigma_{yy} & \tau_{yz} \\ \tau_{zx} & \tau_{zy} & \sigma_{zz} \end{pmatrix} \tag{2.1}$$

where $\tau_{ij} = \tau_{ji}$ and as a rule it is supposed $\sigma_{xx} \equiv \sigma_x$, $\sigma_{yy} \equiv \sigma_y, \sigma_{zz} \equiv \sigma_z$.

There is an interesting interpretation of a tensor as the vector in 9-dimensional space /7/ the part of which is shown in Fig. 1.22.

Tensor (2.1) fully defines the stress state around a point because unit force $\mathbf{p}_n$ at any cross-section through this point with normal $\mathbf{n}$ (n_x, n_y, n_z) can be found from static equations (Fig. 2.2 where stresses from Fig. 2.1 act on the opposite sides of the cube) as

$$\begin{aligned} p_{nx} &= \sigma_x n_x + \tau_{xy} n_y + \tau_{xz} n_z, \\ p_{ny} &= \tau_{xy} n_x + \sigma_y n_y + \tau_{yz} n_z, \\ p_{nz} &= \tau_{xz} n_x + \tau_{yz} n_y + \sigma_z n_z. \end{aligned} \tag{2.2}$$

The projection of $\mathbf{p}_n$ on $\mathbf{n}$ gives normal stress σ_n as

$$\sigma_n = \sigma_x (n_x)^2 + \sigma_y (n_y)^2 + \sigma_z (n_z)^2 + 2\tau_{xy} n_x n_y + 2\tau_{xz} n_x n_z + 2\tau_{yz} n_y n_z.$$

Components τ_{nm} and τ_{nl} can be determined in the same manner. The transformation of all the tensor stresses can be written in the form (i, j = x, y, z; r, s = n, m, l)

$$\sigma_{rs} = \sum_i \sum_j \sigma_{kl} \cos(i,\ r) \cos(j,\ s). \tag{2.3}$$

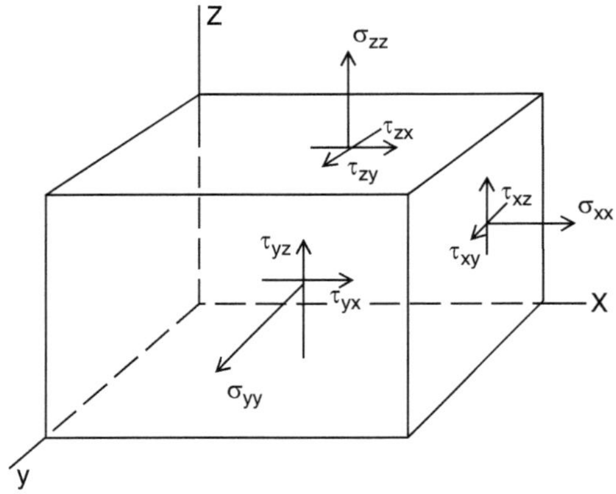

Fig. 2.1. Cube with stresses on its faces

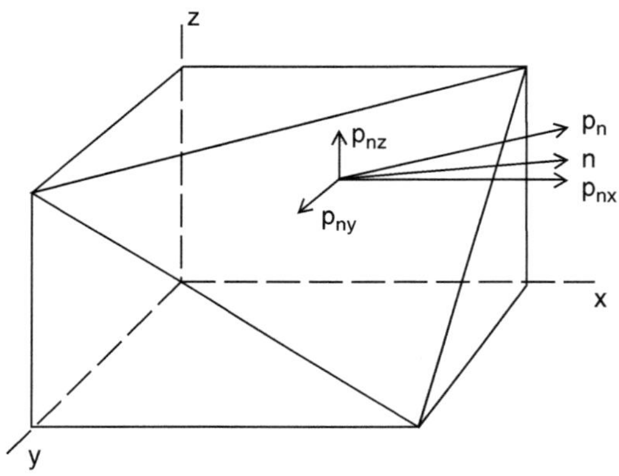

Fig. 2.2. Decomposition of vector $\mathbf{p}_n$

When the cube rotates in a body stresses on its planes change and such a position of it can be found that shearing components become zero. The corresponding faces and normal stresses on them are called main (principal) planes and stresses respectively. They can be found from conditions

$$(n_x)^2 + (n_y)^2 + (n_z)^2 = 1, \tag{2.4}$$

$$\begin{vmatrix} \sigma_x - \sigma_n & \tau_{xy} & \tau_{xz} \\ \tau_{xy} & \sigma_y - \sigma_n & \tau_{yz} \\ \tau_{xz} & \tau_{yz} & \sigma_z - \sigma_n \end{vmatrix} = 0. \tag{2.5}$$

The last of them gives equation relatively to σ_n

$$(\sigma_n)^3 - I_1(\sigma_n)^2 + I_2\sigma_n - I_3 = 0 \tag{2.6}$$

where n = 1, 2, 3, $\sigma_1 \geq \sigma_2 \geq \sigma_3$ – main stresses and I_1, I_2, I_3 – invariants; they do not depend on the position of the cube and are computed as

$$I_1 = \sigma_x + \sigma_y + \sigma_z = \sigma_1 + \sigma_2 + \sigma_3, \quad (2.7)$$
$$I_2 = \sigma_x\sigma_y + \sigma_z\sigma_y + \sigma_x\sigma_z - (\tau_{xy})^2 - (\tau_{xz})^2 - (\tau_{yz})^2 = \sigma_1\sigma_2 + \sigma_2\sigma_3 + \sigma_1\sigma_3, \quad (2.8)$$
$$I_3 = \sigma_x\sigma_y\sigma_z + 2\tau_{xy}\tau_{xz}\tau_{yz} - \sigma_x(\tau_{yz})^2 - \sigma_y(\tau_{xz})^2 - \sigma_z(\tau_{xy})^2 = \sigma_1\sigma_2\sigma_3. \quad (2.9)$$

In the Mechanics of Fracture the fundamental role plays the maximum shearing stress that can be defined as follows

$$\tau_e = 0.5(\sigma_1 - \sigma_3). \quad (2.10)$$

From (2.2) with a help of the Gauss-Ostrogradski's theorem /6/ the differential static equations can be found:

$$\begin{aligned} \partial\sigma_x/\partial x + \partial\tau_{xy}/\partial y + \partial\tau_{xz}/\partial z + X = 0, \\ \partial\tau_{xy}/\partial x + \partial\sigma_y/\partial y + \partial\tau_{yz}/\partial z + Y = 0, \\ \partial\tau_{xz}/\partial x + \partial\tau_{yz}/\partial y + \partial\sigma_z/\partial z + Z = 0 \end{aligned} \quad (2.11)$$

where X, Y, Z are volume's forces. As we can see from (2.11) six stresses are linked only by three expressions and the problem is statically indeterminate. To solve it the strains must be put in consideration.

2.2 Displacements and Strains

External forces deform a body, so its points move in a new position by displacements $\mathbf{u}(u_x,u_y,u_z)$ which determine strains as /5/

$$\varepsilon_x = \partial u_x/\partial x,\ \varepsilon_y = \partial u_y/\partial y,\ \varepsilon_z = \partial u_z/\partial z,$$
$$\gamma_{xy} = \partial u_x/\partial y + \partial u_y/\partial x,\ \gamma_{xz} = \partial u_z/\partial x + \partial u_x/\partial z,\ \gamma_{yz} = \partial u_z/\partial y + \partial u_y/\partial z. \quad (2.12)$$

Here is also valid $\gamma_{ij} = \gamma_{ji}, \varepsilon_{xx} \equiv \varepsilon_x, \varepsilon_{yy} \equiv \varepsilon_y, \varepsilon_{zz} \equiv \varepsilon_z$ and strains form a tensor similar to (2.1) (see errata to 1st ed.)

$$T_\varepsilon = \begin{pmatrix} \varepsilon_x & \gamma_{xy}/2 & \gamma_{xz}/2 \\ \gamma_{xy}/2 & \varepsilon_y & \gamma_{yz}/2 \\ \gamma_{xz}/2 & \gamma_{yz}/2 & \varepsilon_z \end{pmatrix}. \quad (2.13)$$

The strains are linked by well-known compatibility equations

$$\partial^2\varepsilon_x/\partial y^2 + \partial^2\varepsilon_y/\partial x^2 = \partial^2\gamma_{xy}/\partial x\partial y,$$
$$2\partial^2\varepsilon_x/\partial z\partial y = \partial(\partial\gamma_{xy}/\partial z + \partial\gamma_{xz}/\partial y - \partial\gamma_{yz}/\partial x)/\partial x \quad (2.14)$$

and 4 other similar relations can be received by cyclic change of indexes.

The generalization for finite linear strains can be fulfilled with consideration of a change of measured basis (see (1.43)) as

$$\varepsilon_i = \ln(l_i/l_{io}) \quad (i = x, y, z) \quad (2.15)$$

and there is an option for shear /7/.

It is not difficult to notice an analogy between tensors (2.1) and (2.13) where σ_i, τ_{ij} must be replaced by $\varepsilon_i, \gamma_{ij}/2$ and vice versa. In this manner invariants for strains can be found from (2.7)–(2.9) and for maximum shear we have

$$\gamma_m = \varepsilon_1 - \varepsilon_3. \tag{2.16}$$

Now for 15 unknowns we have 9 equations and in order to close the system we must add some other 6 equations. They are constitutive laws which are not the same for different materials.

2.3 Rheological Equations

2.3.1 Generalised Hooke's Law

In a certain range of deformation the Hooke's law can be used. It is usually written in form

$$\begin{aligned} \varepsilon_x &= (\sigma_x - \nu(\sigma_y + \sigma_z))/E, \quad \gamma_{zy} = \tau_{zy}/G, \\ \varepsilon_y &= (\sigma_y - \nu(\sigma_x + \sigma_z))/E, \quad \gamma_{xz} = \tau_{xz}/G \\ \varepsilon_z &= (\sigma_z - \nu(\sigma_x + \sigma_y))/E, \quad \gamma_{xy} = \tau_{xy}/G. \end{aligned} \tag{2.17}$$

where modulus of shear G is linked with E and the Poisson's ratio ν by expression

$$G = E/2(1+\nu). \tag{2.18}$$

In practice other forms of the Hooke's law are often applied. In order to get some of them we summarize three left relations (2.17) and find

$$\sigma_m = Ee_m/3(1-2\nu). \tag{2.19}$$

Here

$$e_m = \varepsilon_x + \varepsilon_y + \varepsilon_z \tag{2.20}$$

– a relative change of materials volume and

$$\sigma_m = (\sigma_x + \sigma_y + \sigma_z)/3 \tag{2.21}$$

is the mean stress. Quantities σ_m, e_m are invariants of respective tensors (see above (2.7) and the analogy between them).

Taking off $e_m/3$ from the left three expressions (2.17) we receive relations

$$e_i = (1+\nu)S_i/E \tag{2.22}$$

where e_i, S_i (i = x, y, z or 1, 2, 3 for main strains and stresses) are components of consequent deviators (tensors with sums of diagonal elements equal to zero).

Considering (2.22), (2.18) and three right equations (2.17) we can represent the Hooke's law as

$$S_{ij} = 2Ge_{ij} \quad (i, j = x, y, z) \tag{2.23}$$

which must be used together with (2.19). Here for $i \neq j$ $S_{ij} \equiv \tau_{ij}, e_{ij} \equiv \gamma_{ij}/2$. There are other forms of the Hooke's law in the literature.

In practice the effective values σ_e, ε_e linked with the second invariants of consequent deviators (for which expression (2.8) is valid with respective replacements) are used. They are usually written as follows (ε_e is given for incompressible body)

$$\sigma_e = \sqrt{1/2}\sqrt{(\sigma_1 - \sigma_2)^2 + (\sigma_2 - \sigma_3)^2 + (\sigma_3 - \sigma_1)^2}, \tag{2.24}$$

$$\varepsilon_e = \left(2/\sqrt{3}\right)\sqrt{\varepsilon_1^2 + \varepsilon_3^2 + \varepsilon_1\varepsilon_3}. \tag{2.25}$$

With consideration of (2.24), (2.25) relations (2.23) can be represented in the form useful for generalizations as

$$e_{ij} = (3\varepsilon_e/2\sigma_e)S_{ij}. \tag{2.26}$$

2.3.2 Non-linear Equations

In the theory of ideal plasticity expressions (2.26) mean only a proportionality of the deviators since ratio ε_e/σ_e is taken constant. Above that condition

$$\sigma_e = \sigma_{yi}(\tau_e = \tau_{yi} = \sigma_{yi}/2), \tag{2.27}$$

where σ_{yi} was given in Chap. 1 and τ_{yi} is an yielding point in shear, together with demand of the constant material's volume is used.

For a hardening plastic body a dependence of ε_e on σ_e must be known and (2.26) becomes

$$e_{ij} = (3\varepsilon_e(\sigma_e)/2\sigma_e)S_{ij}. \tag{2.28}$$

The power law for this purpose is often applied in form

$$\varepsilon_e = \Omega(\sigma_e)^m \tag{2.29}$$

Here Ω, m are constants which can be found from tests in tension, compression or bending (see Chap. 1). Here below expression (2.29) will be also used as a dependence of stresses on strains:

$$\sigma_e = \omega(\varepsilon_e)^\mu \tag{2.30}$$

where $\omega = \Omega^{-\mu}(1 \geq \mu \geq 0)$.

Replacing in (2.28) strains e_{ij} by their rates we get a power law for a steady creep (see also (1.41))

$$de_{ij}/dt = (3B/2)(\sigma_{eq})^{m-1}S_{ij}. \tag{2.31}$$

Here σ_{eq} is an equivalent stress that takes into account an influence of a stress state type on curves in coordinates σ_e, ε_e.

The simplest way to describe an unsteady creep with damage is to generalize relation (1.45) (for incompressible body) as /7/

$$\exp(-\alpha\varepsilon_{eq})\varepsilon_{ij} = (3\Omega(t)/2)(\sigma_{eq})^{m-1}S_{ij}. \tag{2.32}$$

where ε_{eq} – an equivalent strain which determines a development of a damage. The experiments /14/ show that as ε_{eq} the biggest main strain ε_1 may be taken. If the influence of time is small (2.32) becomes the plasticity law (2.28) generalized here.

2.3.3 Constitutive Equations for Anisotropic Materials

Equation (2.32) can be generalized for an anisotropic body /7/. If main axes of stress state coincide with orthotropic directions we have

$$\varepsilon_i\exp(-\alpha\varepsilon_{eq}) = \Omega(t)(\sigma_{eq})^{m-1}(k_j(\sigma_i - \sigma_k) + k_k(\sigma_i - \sigma_j)). \tag{2.33}$$

Here k_s (s = i, j, k or x, y, z) are anisotropy parameters. Their role can be appreciated by a divergence of curves in coordinates σ_e, ε_e (scalar properties of a material) and deviations from the condition $\mu_\varepsilon = \mu_\sigma$ (its vector properties) where $\mu_\varepsilon, \mu_\sigma$ are the Lode's parameters (they are linked with the third invariants of respective deviators) which are determined by well-known relations /22/

$$\mu_\sigma = (2\sigma_2 - \sigma_1 - \sigma_3)/(\sigma_1 - \sigma_3) \quad (1 \geq \mu_\sigma \geq -1), \tag{2.34}$$

$$\mu_\varepsilon = 3\varepsilon_2/(\varepsilon_1 - \varepsilon_3) \quad (1 \geq \mu_\varepsilon \geq -1) \tag{2.35}$$

where the latter expression is given for an incompressible body.

Soils are usually stratified or laminated in horizontal directions. For this case we can use a model of transversally isotropic body with vertical symmetry axis z (Fig. 2.3).Putting into (2.33) $k_i = k_j = 0.5$ we have after transformations

$$\begin{aligned} \exp(-\alpha\varepsilon_{eq})\varepsilon_i &= \Omega_1(t)(\sigma_{eq})^{m-1}(\sigma_i - k\sigma_j - (1-k)\,\sigma_k), \\ \exp(-\alpha\varepsilon_{eq})\varepsilon_k &= \Omega_1(t)(\sigma_{eq})^{m-1}(2\sigma_k - \sigma_i - \sigma_j)(1-k). \end{aligned} \tag{2.36}$$

where i, j = x, y; $2k_k = k/(1-k)$, $\Omega_1 = \Omega/2(1-k)$ and the value of k can be found from the results of tension or compression in i-direction at $\sigma_j = \sigma_k = 0$ according to relation $k = -\varepsilon_j/\varepsilon_i$.

When k = 0.5 the material is isotropic. If k = 1 it can be modelled as a system of fibres parallel to axis k in a feeble matrix (a wood, for example). Cases k = 0 and k = −1 can be interpreted as systems of unconnected bars

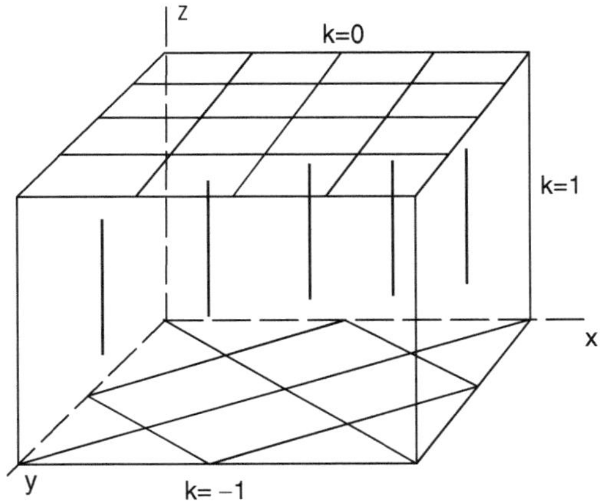

Fig. 2.3. Cube with axis of symmetry z

in directions of axes i, j or under angles $\pi/4$ to them respectively (Fig. 2.3). The last two models correspond to a play-wood.

In the case of plane deformation we have from (2.36) at $\varepsilon_y = 0$. $\sigma_y = k\sigma_x + (1-k)\sigma_z$ and hence

$$\exp(-\alpha\varepsilon_{eq})\varepsilon_z = -\exp(-\alpha\varepsilon_{eq})\varepsilon_x = \Omega_1(t)(\sigma_{eq})^{m-1}(\sigma_z - \sigma_x)(1-k^2). \quad (2.37)$$

Similar expressions can be found from (2.33) at if $k_x = k_y = 0.5$. We can see that law (2.37) differs from (2.32) by a constant multiplier.

Now we study the rheological properties of a transversally isotropic material in a plane stress state for three options of interposition of loading and isotropy planes. Using relations (2.24), (2.25) and (2.36) we receive

$$\exp(-\alpha\varepsilon_{eq})\varepsilon_e = 2\Omega_1(3(1-n+n^2))^{-1/2}\sqrt{\Lambda}(\sigma_{eq})^{m-1}\sigma_e \quad (2.38)$$

where $n = \sigma_y/\sigma_x$ and the structure of function $\Lambda(k, n)$ depends on a mutual position of symmetry and loading planes.

For the case of Fig. 2.3 at $\sigma_k = 0$ we have from the first expression (2.36)

$$\exp(-\alpha\varepsilon_{eq})\varepsilon_i = \Omega_1(t)(\sigma_{eq})^{m-1}(\sigma_i - k\sigma_j) \quad (i, j, = x\ y). \quad (2.39)$$

In this option

$$\Lambda = (1-k+k^2)(1+n^2) + n(k^2 - 4k + 1). \quad (2.40)$$

Curves $\sigma_e(\varepsilon_e)$ according to (2.38) at $\alpha = 0$, (2.40) are represented in Fig. 2.4 by solid ($k = 0.5$), broken ($k = 1$), interrupted by points ($k = 0$) and dotted ($k = -1$) lines for n equal to 0 and 1. It is easy to see that the body $k = 1$ has an absolute rigidity at $n = 1$ and small resistance to deformation at $n = 0$. This result is well-known from compression tests of timber along and

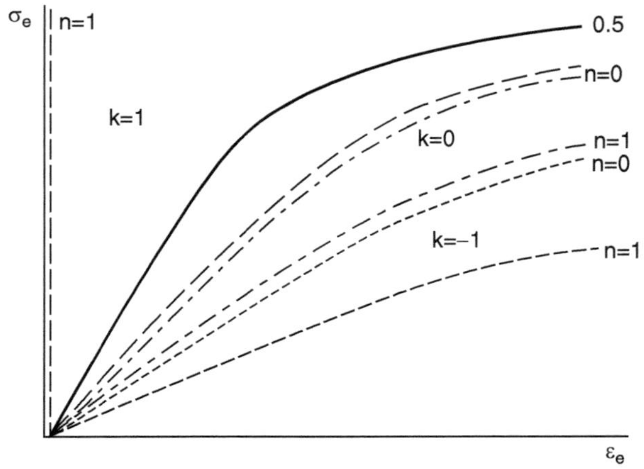

Fig. 2.4. Curves $\sigma_e(\varepsilon_e)$ at different n and k when z is axis of symmetry

transverse the fibres respectively. We must also notice that when parameter k diminishes the material's rigidity falls.

According to (2.34), (2.35) we find for the cases $\varepsilon_2 = \varepsilon_y$ and $\varepsilon_2 = \varepsilon_z$ respectively

$$\mu_\varepsilon = 3(\mu_\sigma \pm (1 - 2k))/(5 - 4k \pm (1 - 2k)\mu_\sigma), \quad (2.41)$$

$$\mu_\varepsilon = 3(1 - k)(3 \pm \mu_\sigma)/(1 + k)(\mu_\sigma \pm (-1)) \quad (2.42)$$

at upper signs. The corresponding curves are shown by solid and broken lines in Fig. 2.5.

From this picture we can see that the influence of k on the vector properties of the body is also high and condition $\mu_\varepsilon = \mu_\sigma$ fulfils for an isotropic body and in relation (2.42) at $\mu_\varepsilon = \mu_\sigma = -1$.

A similar situation takes place in two other options of isotropy and loading planes interposition when we derive from (2.36)

$$\begin{aligned} \exp(-\alpha\varepsilon_{eq})\varepsilon_i &= \Omega_1(t)(\sigma_{eq})^{m-1}(1 - k)(2\sigma_i - \sigma_j), \\ \exp(-\alpha\varepsilon_{eq})\varepsilon_j &= \Omega_1(t)(\sigma_{eq})^{m-1}(\sigma_j - (1 - k)\sigma_i). \end{aligned} \quad (2.43)$$

In (2.43) parameter k can be determined according to relation $k = -\varepsilon_z/\varepsilon_j$ at $\sigma_i = 0$ where j is axis of symmetry.

If $i = x$ we have for Λ

$$\Lambda = 3(1 - k)^2(1 - n) + n^2(1 - k + k^2) \quad (2.44)$$

and function $\sigma_e(\varepsilon_e)$ can be constructed as in Fig. 2.4 with the mutual replacement of curves $n = 0$ and $n = 1$. Expressions (2.41), (2.42) for $\varepsilon_2 = \varepsilon_y$ and $\varepsilon_2 = \varepsilon_x$ can be used at lower signs and consequent diagram $\mu_\varepsilon(\mu_\sigma)$ can be found from Fig. 2.5 by rotating it about the centre on angle π.

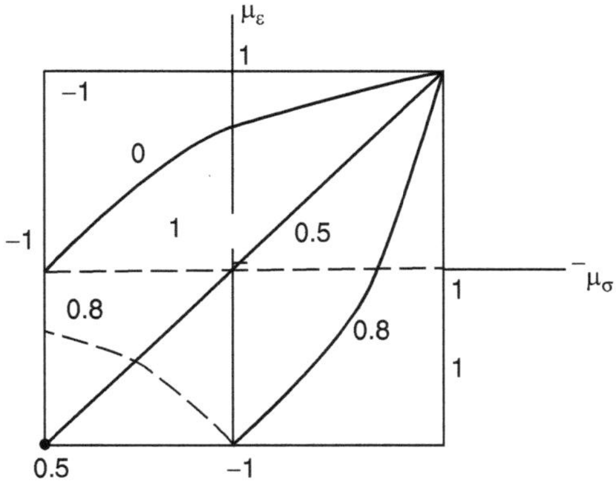

Fig. 2.5. Dependence $\mu_\varepsilon(\mu_\sigma)$ at different k when z is axis of symmetry

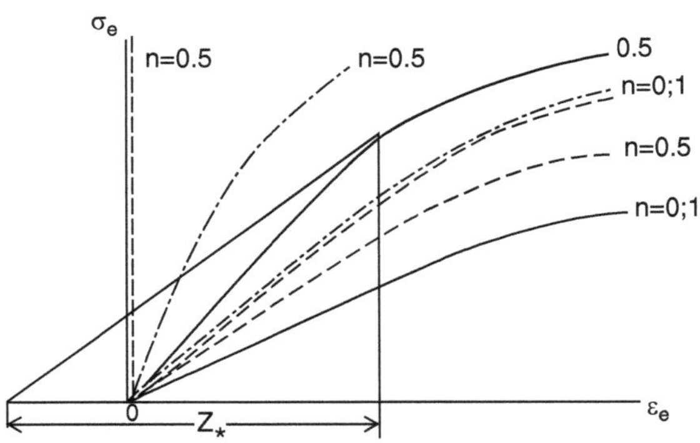

Fig. 2.6. Curves $\sigma_e(\varepsilon_e)$ at different n and k when y is axis of symmetry

Such similarities of these two cases can be explained by the same position of the plane with τ_e relatively to the symmetry axis.

If in (2.43) i = y is the axis of symmetry function $\Lambda(n, k)$ can be found by replacement in (2.38), (2.44) n by 1/n. Corresponding curves $\sigma_e(\varepsilon_e)$ are given in Fig. 2.6 by the same lines as in Fig. 2.4. In order to find all expressions $\mu_\varepsilon(\mu_\sigma)$ we must consider three possibilities of relations between strains when we have

$$\mu_\varepsilon = 3\mu_\sigma(1 - k)/(1 + k), \tag{2.45}$$

$$\mu_\varepsilon = 3(\pm(-(1 + k)) - \mu_\sigma(1 - k))/(1 + k \pm \mu_\sigma(k - 1)) \tag{2.46}$$

for the options $\varepsilon_2 = \varepsilon_y, \varepsilon_2 = \varepsilon_z$ (upper signs) and $\varepsilon_2 = \varepsilon_x$ (lower signs) respectively. The corresponding diagrams are given by solid, broken and interrupted by points lines in Fig. 2.7 and we can see also the high influence of k here Option (2.45) embraces the first and the third quadrants of the plane when k changes from −1 to 1 including straight line k = 0.5.

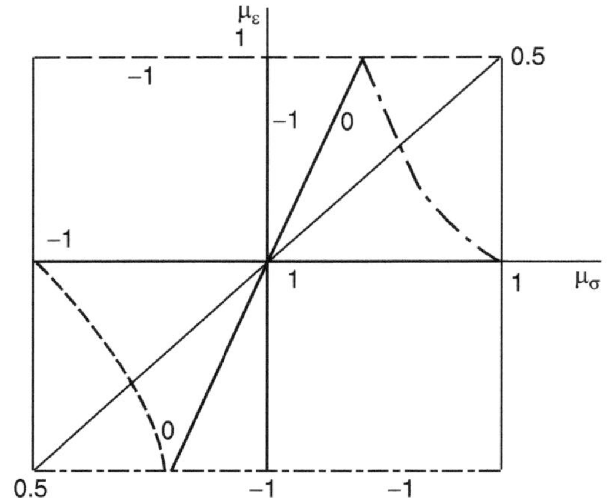

Fig. 2.7. Function $\mu_\varepsilon(\mu_\sigma)$ at different k when y is axis of symmetry

Following the idea of (2.33) we can generalize (2.31) for a steady creep in a similar way as

$$d\varepsilon_i/dt = B(\sigma_{eq})^{m-1}(k_j(\sigma_i - \sigma_k) + k_k(\sigma_i - \sigma_j)). \tag{2.47}$$

and the option of transversally isotropic material can be also developed.

2.4 Solution Methods of Mechanical Problems

2.4.1 General Considerations

In the common case of the continuum mechanics it is necessary to find 15 functions from 15 equations ((2.11), (2.12) and of (2.17) type). Searched variables should satisfy boundary conditions when stresses or strains are given on parts of body's surface. The consequent problems are called the first and the second border tasks. A mixed problem appears when stresses and strains are given simultaneously on different parts of the surface.

The usual way to solve the problem consists in exclusion of variables, so only stresses or strains (displacements) remain. But up to date no final results are known for the general task even at ideal materials in consideration. For this reason simplifying hypotheses of geometrical character are introduced when the task is studied as the anti-plane (a longitudinal shear), plane, axisymmetric etc. problem.

2.4.2 Basic Equations for Anti-plane Deformation

In this case we have only five unknowns (Fig. 2.8): $\tau_{xz} \equiv \tau_x, \tau_{yz} \equiv \tau_y, \gamma_{xz} \equiv \gamma_x, \gamma_{yz} \equiv \gamma_y$ and u_z.

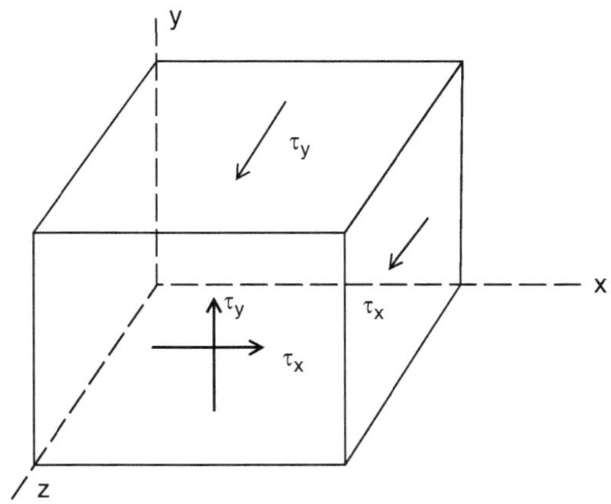

Fig. 2.8. Stresses at anti-plane deformation

According to (2.11)–(2.14) they are linked by equations

$$\partial\tau_x/\partial x + \partial\tau_y/\partial y = 0, \tag{2.48}$$

$$\gamma_x = \partial u_z/\partial x,\ \gamma_y = \partial u_z/\partial y, \tag{2.49}$$

$$\partial\gamma_x/\partial y - \partial\gamma_y/\partial x = 0. \tag{2.50}$$

In elastic range we receive from (2.15)

$$\tau_x = G\gamma_x,\ \tau_y = G\gamma_y \tag{2.51}$$

and for 5 unknowns we have 5 equations (2.48), (2.49) and (2.51) or for 4 variables τ_x, τ_y, γ_x, γ_y – 4 equations: (2.48), (2.50) and (2.51).

For an ideal plastic body condition $\tau_e = \tau_{yi}$ where

$$\tau_e = \sqrt{(\tau_x)^2 + (\tau_y)^2} \tag{2.52}$$

makes the problem statically determinate.

Lastly at unsteady creep we have from (2.32)

$$\exp(-\alpha\varepsilon_{eq})\gamma_i = \Omega(t)(\tau_e)^{m-1}\tau_i \tag{2.53}$$

where i = x, y and maximum shear may be computed through γ_x, γ_y by expression similar to (2.52):

$$\gamma_m = \sqrt{(\gamma_x)^2 + (\gamma_y)^2}. \tag{2.54}$$

For some problems it is more convenient to use polar coordinates $r = \sqrt{x^2 + y^2}$, $\tan^{-1}\theta = y/x$ (Fig. 2.9) where

$$\tau_r \equiv \tau_{rz} = \tau_x \cos\theta + \tau_y \sin\theta,\ \tau_\theta \equiv \tau_{\theta z} = -\tau_x \sin\theta + \tau_y \cos\theta \tag{2.55}$$

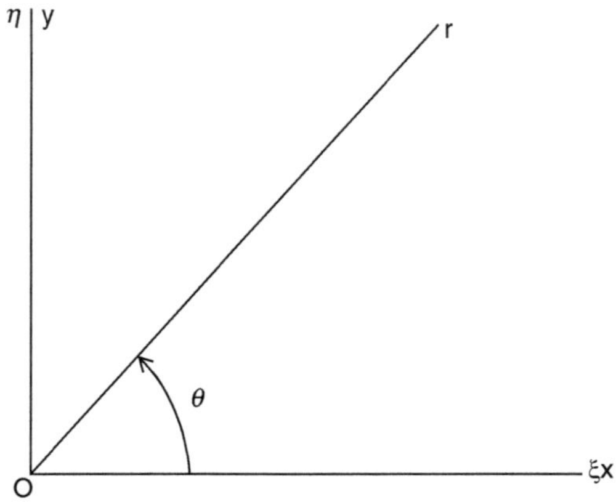

Fig. 2.9. Polar coordinate system

and instead of (2.48) we have

$$\partial(\mathrm{r}\tau_{\mathrm{r}})/\partial\mathrm{r} + \partial\tau_\theta/\partial\theta = 0. \tag{2.56}$$

The shearing strains in this case are:

$$\gamma_{\mathrm{r}} \equiv \gamma_{\mathrm{rz}} = \partial\mathrm{u_z}/\partial\mathrm{r},\ \gamma_\theta \equiv \gamma_{\theta\mathrm{z}} = \partial\mathrm{u_z}/\mathrm{r}\partial\theta \tag{2.57}$$

and compatibility law (2.50) becomes

$$\partial\gamma_{\mathrm{r}}/\partial\theta - \partial(\mathrm{r}\gamma_\theta)/\partial\mathrm{r} = 0. \tag{2.58}$$

Expressions (2.51)–(2.54) are valid at the replacement in them x, y by r, θ.

2.4.3 Plane Problem

Here 8 main variables do not depend on one coordinate (for example z) and for them we have from (2.11), (2.12) and (2.14) common equations

$$\partial\sigma_{\mathrm{x}}/\partial\mathrm{x} + \partial\tau_{\mathrm{xy}}/\partial\mathrm{y} + \mathrm{X} = 0,\ \partial\tau_{\mathrm{xy}}/\partial\mathrm{x} + \partial\sigma_{\mathrm{y}}/\partial\mathrm{y} + \mathrm{Y} = 0, \tag{2.59}$$

$$\varepsilon_{\mathrm{x}} = \partial\mathrm{u_x}/\partial\mathrm{x},\ \varepsilon_{\mathrm{y}} = \partial\mathrm{u_y}/\partial\mathrm{y},\ \gamma_{\mathrm{xy}} = \partial\mathrm{u_x}/\partial\mathrm{y} + \partial\mathrm{u_y}/\partial\mathrm{x}, \tag{2.60}$$

$$\partial^2\varepsilon_{\mathrm{x}}/\partial\mathrm{y}^2 + \partial^2\varepsilon_{\mathrm{y}}/\partial\mathrm{x}^2 = \partial^2\gamma_{\mathrm{xy}}/\partial\mathrm{x}\partial\mathrm{y}. \tag{2.61}$$

The Hooke's law (2.17) in the case $\sigma_z = 0$ (plane stress state) becomes

$$\mathrm{E}\varepsilon_{\mathrm{x}} = \sigma_{\mathrm{x}} - \nu\sigma_{\mathrm{y}},\ \mathrm{E}\varepsilon_{\mathrm{y}} = \sigma_{\mathrm{y}} - \nu\sigma_{\mathrm{x}},\ \mathrm{G}\gamma_{\mathrm{xy}} = \tau_{\mathrm{xy}}. \tag{2.62}$$

If $\varepsilon_z = 0$ (plane deformation) then we find from the third expression (2.17) $\sigma_z = \nu(\sigma_x + \sigma_y)$ and relations (2.62) are valid at the replacement in them E,

ν by $E/(1-\nu^2)$, $\nu/(1-\nu)$ respectively and it is not difficult to prove that G-value remains the same.

Excluding from (2.59), (2.60), (2.62) stresses and strains we get two equations for u_x, u_y. If we solve the problem in stresses we can receive from (2.59)–(2.62) a byharmonic equation for a function Φ as

$$\partial^4\Phi/\partial x^4 + 2\partial^4\Phi/\partial x^2\partial y^2 + \partial^4\Phi/\partial y^4 = 0 \tag{2.63}$$

where

$$\sigma_x = \partial^2\Phi/\partial y^2,\ \sigma_y = \partial^2\Phi/\partial x^2,\ \tau_{xy} = -\partial^2\Phi/\partial x\partial y - yX - xY. \tag{2.64}$$

The maximum shearing stress in this problem is

$$\tau_e = 0.5\sqrt{(\sigma_x - \sigma_y)^2 + 4(\tau_{xy})^2} \tag{2.65}$$

and here we can also conclude that for ideal plasticity condition $\tau_e = \tau_{yi} = \sigma_{yi}/2$ together with (2.59) makes the problem statically determinate one. For hardening at creep body we have for the plane deformation

$$\begin{aligned} \exp(-\alpha\varepsilon_{eq})\varepsilon_x &= -\exp(-\alpha\varepsilon_{eq})\varepsilon_y = 3\Omega(t)(\sigma_{eq})^{m-1}(\sigma_x - \sigma_y)/4,\\ \exp(-\alpha\varepsilon_{eq})\gamma_{xy} &= 3\Omega(\sigma_{eq})^{m-1}\tau_{xy}. \end{aligned} \tag{2.66}$$

The main equations in the polar coordinate system (Fig. 2.9) can be derived directly or by transformations of corresponding expressions in Decart's variables x, y. In /7/ a simplified procedure is used for this purpose (after differentiation we equal θ to zero). As a result we receive firstly static laws as

$$r\partial\sigma_r/\partial r + \sigma_r - \sigma_\theta + \partial\tau_{r\theta}/\partial\theta = 0,\ \partial(r^2\tau_{r\theta})/\partial r + r\partial\sigma_\theta/\partial\theta = 0. \tag{2.67}$$

Combining (2.67) we receive a very useful expression

$$\partial^2\tau_{r\theta}/\partial\theta^2 + \partial(r\partial(\sigma_r - \sigma_\theta)/\partial\theta)/\partial r - r\partial(\partial(r^2\tau_{r\theta})/r\partial r)/\partial r = 0 \tag{2.68}$$

Strains are linked with displacements by relations similar to (2.60) and they can be received from them with the help of vector transformation equations similar to (2.55):

$$\varepsilon_r = \partial u_r/\partial r,\ \varepsilon_\theta = u_r/r + \partial u_\theta/r\partial\theta,\ \gamma_{r\theta} = r\partial(u_\theta/r)/\partial r + \partial u_r/r\partial\theta. \tag{2.69}$$

Condition of constant volume $\varepsilon_r + \varepsilon_\theta = 0$ can be written with the help of two first expressions (2.69) as

$$\partial(ru_r)/\partial r = -\partial u_\theta/\partial\theta \tag{2.70}$$

and instead of (2.61) we have

$$r\partial(\partial(r^2\varepsilon_\theta)/r\partial r)/\partial r - \partial^2\varepsilon_\theta/\partial\theta^2 = \partial^2(r\gamma_{r\theta})/\partial r\partial\theta. \tag{2.71}$$

It is valid to notice the identity of (2.71) and (2.68) if we replace ε_θ, $\gamma_{r\theta}$ by $\tau_{r\theta}$ and $\sigma_r - \sigma_\theta$ respectively.

The Hooke's law (2.62), relations (2.65), (2.66) are valid here with the replacement in them x, y by r, θ. Transformation equations (2.3) can be expressed here as (see also Fig. 2.9)

$$\begin{aligned}
\tau_{r\theta} &= 0.5(\sigma_y - \sigma_x)\sin 2\theta + \tau_{xy}\cos 2\theta,\\
\sigma_r &= \sigma_m + 0.5(\sigma_x - \sigma_y)\cos 2\theta + \tau_{xy}\sin 2\theta,\\
\sigma_\theta &= \sigma_m - 0.5(\sigma_x - \sigma_y)\cos 2\theta - \tau_{xy}\sin 2\theta
\end{aligned} \tag{2.72}$$

$$\begin{matrix}\sigma_1\\ \sigma_3\end{matrix} = \sigma \pm \tau_e \tag{2.73}$$

where (see the Mohr's circle in Fig. 1.22)

$$\sigma_m = 0.5(\sigma_x + \sigma_y) = 0.5(\sigma_\theta + \sigma_r).$$

Lastly instead of (2.63), (2.64) we write

$$(\partial/r\partial r + \partial^2/r^2\partial\theta^2 + \partial^2/\partial r^2)(\partial\Phi/r\partial r + \partial^2\Phi/r^2\partial\theta^2 + \partial^2\Phi/\partial r^2) = 0, \tag{2.74}$$

$$\sigma_r = \partial\Phi/r\partial r + \partial^2\Phi/r^2\partial\theta^2,\ \sigma_\theta = \partial^2\Phi/\partial r^2,\ \tau = -\,\partial(\partial\Phi/r\partial\theta)\partial r. \tag{2.75}$$

2.4.4 Axisymmetric Problem

In this case cylindrical coordinates (Fig. 2.10) can be used and if all the variables do not depend on angle θ the main equations are:

$$\begin{aligned}
&\varepsilon_r = \partial u_r/\partial r,\ \varepsilon_\theta = u_r/r,\ \varepsilon_z = \partial u_z/\partial z,\ \gamma_{rz} = \partial u_r/\partial z + \partial u_z/\partial r,\\
&\partial\sigma_r/\partial r + \partial\tau_{rz}/\partial z + (\sigma_r - \sigma_\theta)/r = 0,\ \partial\tau_{rz}/\partial r + \partial\sigma_z/\partial z + \tau_{rz}/r = 0,\\
&\Delta\sigma_r - 2(\sigma_r - \sigma_\theta)/r^2 + 3(\partial^2\sigma_r/\partial r^2)/(1+\nu) = 0,\\
&\Delta\sigma_\theta + 2(\sigma_r - \sigma_\theta)/r^2 + 3(\partial\sigma_r/r\partial r)/(1+\nu) = 0.
\end{aligned} \tag{2.76}$$

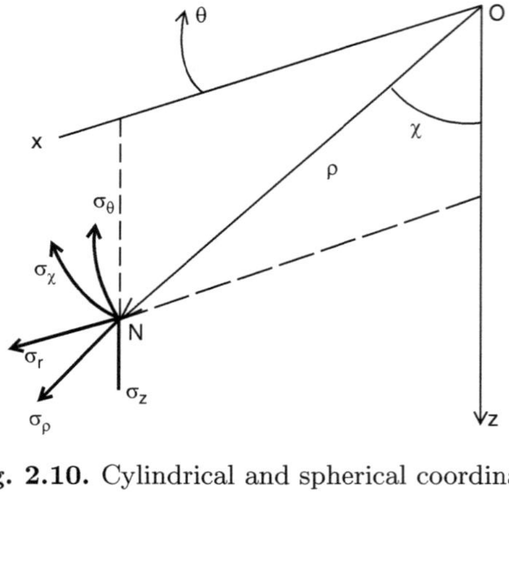

Fig. 2.10. Cylindrical and spherical coordinates

The first line σ_χ in (2.76) is the Saint-Venant's expressions, the second one represents (2.78) $-\rho(\partial(\rho^3\tau_{\rho\chi})/\rho^2\partial\rho)/\partial\rho$ static law and the latter two are compatibility equations in which Δ is the Laplace's operator of this task

$$\Delta = \partial^2/\partial r^2 + r^{-1}\partial/\partial r + \partial^2/\partial z^2.$$

2.4.5 Spherical Coordinates

In the option when all the variables are independent on θ the initial static laws can be written as

$$\begin{aligned} &\rho\partial\sigma_\rho/\partial\rho + \partial\sigma_\chi/\partial\chi + 2\sigma_\rho - \sigma_\chi - \sigma_\theta + \tau_{\rho\chi}\cot\chi = 0, \\ &\partial(\rho^3\tau_{\rho\chi})/\rho^2\partial\rho + \partial\sigma_\chi/\partial\chi + (\sigma_\theta - \sigma_\chi)\cot\chi = 0. \end{aligned} \tag{2.77}$$

Combining (2.77) at $\sigma_\theta = \sigma_\chi$ we receive a very useful equation

$$\begin{aligned} &\partial^2(\rho^2(\sigma_\rho - \sigma_\chi))/\rho\partial\rho\partial\chi + \partial^2\tau_{\rho\chi}/\partial\chi^2 + \partial(\tau_{\rho\chi}\cot\chi)/\partial\chi \\ &\quad - \rho(\partial(\rho^3\tau_{\rho\chi})/\rho^2\partial\rho)\partial\rho = 0. \end{aligned} \tag{2.78}$$

Strains are linked with displacements by expressions

$$\begin{aligned} \varepsilon_\rho &= \partial u_\rho/\partial\rho,\ \varepsilon_\chi = \partial u_\chi/\rho\partial\chi + u_\rho/\rho,\ \varepsilon_\theta = u_\rho/\rho + u_\chi(\cot\chi)/\rho, \\ \gamma_{\rho\chi} &= \partial u_\rho/\rho\partial\chi + \rho\partial(u_\chi/\rho)/\partial\rho. \end{aligned} \tag{2.79}$$

Three of them submit to compatibility equation

$$\rho\partial(\rho\partial\varepsilon_\chi/\partial\rho)/\partial\rho + \partial^2\varepsilon_\rho/\partial\chi^2 + \rho\partial(\varepsilon_\chi - \varepsilon_\rho)/\partial\rho - \partial^2(\rho\gamma_{\rho\chi})/\partial\chi\partial\rho = 0$$

which at replacement ρ, χ by r, θ gives (2.71). The rheological laws will be given for a concrete task.

If the main variables do not depend on angle χ the equations are simpler and can be written as follows

$$d(\rho^2\sigma_\rho)/\rho d\rho - 2\sigma_\theta = 0,\ \varepsilon_\rho = du_\rho/d\rho,\ \varepsilon_\theta = \varepsilon_\chi = u_\rho/\rho. \tag{2.80}$$

The strains are linked by the conditions of compatibility and of a constant volume:

$$d\varepsilon_\theta/d\rho + (\varepsilon_\theta - \varepsilon_\rho)/\rho = 0,\ d(\rho^2 u_\rho)/d\rho = 0. \tag{2.81}$$

When $u_\theta = u_\chi = 0$ the second and the third equation (2.80) are valid. Relation for $\gamma_{\rho\chi}$ as well as the compatibility law are

$$\gamma_{\rho\chi} = du_\rho/\rho d\chi,\ d\varepsilon_\chi/d\chi = \gamma_{\rho\chi}. \tag{2.82}$$

2.5 Economical Profile of Triangular Dam Under Self-Weight and Lateral Pressure of Water

Brittle materials badly resist tensile stress. Because of that in made of them structures such as triangular dam (Fig. 2.11) only compressive pressures are allowed. In economical profile on face OA the condition $\sigma_y = 0$ is taken and according to the hypothesis of plane horizontal cross-sections $\varepsilon_y = kx$ where k is angle of their rotation we have from (1.48) (see diagram in Fig. 2.11)

$$\sigma_y = \omega(t)k^\mu x^\mu. \tag{2.83}$$

Using integral static equations $\Sigma Y = 0$ and $\Sigma M_A = 0$ we receive

$$-\delta_m y^2 \tan\beta = 2\omega k^\mu \int_0^{y\tan\beta} x^\mu dx,$$

$$-\delta_m y^3 \tan^2\beta - \delta_w y^3 = \omega k^\mu \int_0^{y\tan\beta} x^{\mu+1} dx.$$

Here β is angle in apex of the profile, δ_m, δ_w – specific weights of dam material and water. Computing the integrals we obtain after cancellations

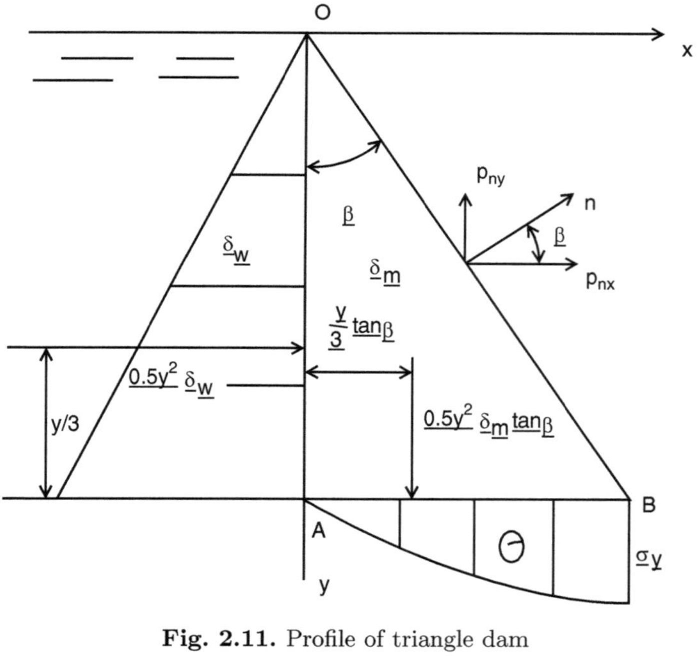

Fig. 2.11. Profile of triangle dam

$$-\delta_m = 2\omega k^\mu y^{\mu-1}(\tan^\mu\beta)/(\mu+1),$$
$$-(\delta_m \tan^2\beta + \delta_w) = 6\omega k^\mu y^{\mu-1}(\tan^{\mu+2}\beta)/(\mu+2). \tag{2.84}$$

Excluding from (2.84) ωk^μ we find minimum angle β for economical profile as

$$\tan\beta = \sqrt{(2+\mu)/(2\mu+1)}\sqrt{\delta_w/\delta_m}. \tag{2.85}$$

At $\mu = 1$ we deduce from (2.85) the well-known elastic solution /51/.

An absence of tensile stresses does not exclude a fracture due to maximum strain. We shall determine it on faces OA and OB. In the first case we have $\sigma_y = \tau_{xy} = 0$ and

$$\sigma_x = q = -y\delta_w \tag{2.86}$$

or with consideration of (2.85)

$$q = -y\delta_m((2\mu+1)/(\mu+2))\tan^2\beta. \tag{2.87}$$

In order to receive main compressive stress on face OB we exclude from (2.83) and the first (2.84) (the second one coincides with it due to (2.85)) multiplier ωk^μ which gives

$$\sigma_y = -0.5\delta_m y(\mu+1). \tag{2.88}$$

Now we use two first equations (2.2) with $\tau_{xz} = \tau_{yz} = 0, p_{nx} = p_{ny} = 0$ in form

$$\sigma_x \cos\beta - \tau_{xy}\sin\beta = 0,\ \tau_{xy}\cos\beta - \sigma_y \sin\beta = 0 \tag{2.89}$$

and obtain

$$\tau_{xy} = -0.5y\delta_m(\mu+1)\tan\beta,\ \sigma_x = -0.5\delta_m y(\mu+1)\tan^2\beta. \tag{2.90}$$

From relations (2.73) for main (principal) stresses (see e.g./47/) we compute $\sigma_1 = 0$ and the stress along line OB as

$$\sigma_3 = -0.5y\delta_m(\mu+1)/\cos^2\beta. \tag{2.91}$$

To find dangerous points we combine from (2.87), (2.91) ratio

$$q/\sigma_3 = 2((2\mu+1)/(2+3\mu+\mu^2))/\sin^2\beta \tag{2.92}$$

At $\mu = 1, \mu = 0$ we have $q = /\sigma_3/\sin^2\beta$ and, so, $/\sigma_3/ > q$. Maximum of ratio (2.92) takes place at $\mu = 0.366$ and is equal to $1.072\sin^2\beta$ or

$$\max(q/\sigma_3) = 1.464/(\delta_m/\delta_w + 1.366).$$

For real materials it is also less than unity. Since $/\sigma_3/$ increases with y the fracture of the economical profile may be expected in the lowest part of face OB. For elastic materials we can find with the help of (2.91)

$$\max\varepsilon = (\nu/E)y_{max}/\cos^2\beta.$$

The use of infinite elongations rate criterion gives for incompressible materials (clay e.g.)

$$\varepsilon_* = 1/\alpha,\ \Omega(t_*) = (\max/\sigma_3/)^m/\alpha e. \tag{2.93}$$

3

Some Elastic Solutions

3.1 Longitudinal Shear

3.1.1 General Considerations

If a force Q acts along axis z (Fig. 3.1) we have from (2.56) a simple equation

$$d(r\tau_r)/dr = 0 \tag{3.1}$$

with an obvious solution $\tau_r = C/r$ where $C = -Q/2\pi$ from the equilibrium equation $\Sigma Z = 0$ of the part of the cylinder in Fig. 3.1. Using the Hooke's law (2.51) and the first expression (2.57) we can write the final result as follows

$$\tau_r = -Q/2\pi r,\ u_z = -(Q/2\pi G)\ln r + u_o \tag{3.2}$$

where u_o is a constant.

Now we study the problem in complex variables $z = x + iy$, $\zeta = \xi + i\eta$ etc. The convenience of such an approach follows from the fact that main unknowns (displacements, stresses and, hence, strains) can be determined by complex variables and their derivatives. According to definition differentiation of a function

$$w(z) = \varphi(x, y) + i\psi(x, y) \tag{3.3}$$

is possible when its real (Re) φ and imaginary (Im) ψ parts satisfy well-known equations of Cauchy-Riemann

$$\partial\varphi/\partial x = \partial\psi/\partial y,\ \partial\varphi/\partial y = -\partial\psi/\partial x. \tag{3.4}$$

If the function w(z) is known the main variables of the task can be found by the following relations (we can prove the second of them using expressions (3.4), (2.48), (2.50), (2.51))

$$u_z = \varphi(x, y)/G = \mathrm{Re}(w(z))/G,\ \tau_x - i\tau_y = (\tau_r - i\tau_\theta)e^{-i\theta} = w'(z). \tag{3.5}$$

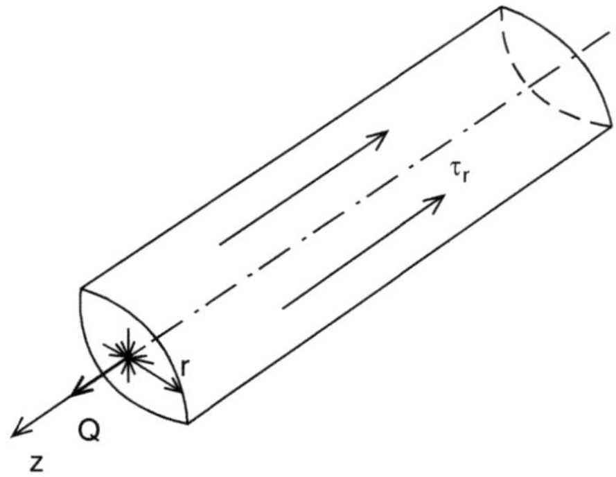

Fig. 3.1. Anti-plane deformation of cylinder

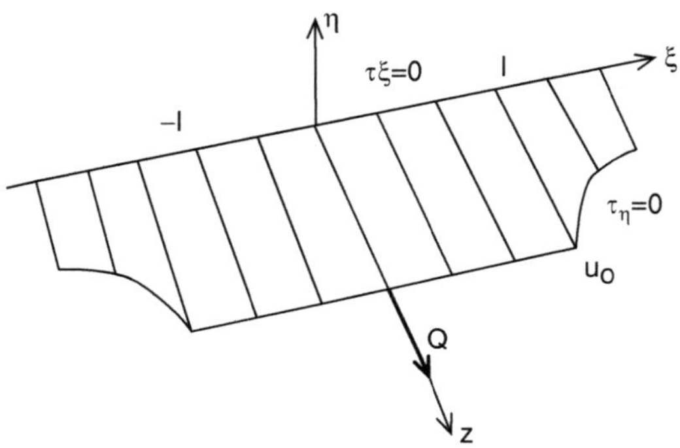

Fig. 3.2. Displacement of strip

The latter equation is derived with the help of relation for vector components transformation (2.55) and sign 'means a derivative by z. We can check expressions above on the example of Fig. 3.1 for which the solution can be given in the following way:

$$w(z) = -(Q/2\pi)\ln z + Gu_o \tag{3.6}$$

The convenience of the complex variables usage consists in the opportunity of conformal transformations application when solutions for simple figures (a semi-plane or a circle) can be transformed to compound sections /16/.

3.1.2 Longitudinal Displacement of Strip

To derive the solution of this problem by the conformal transformation of result (3.6) to the straight line −l, l (Fig. 3.2) with the help of the Zhoukovski's relation which was deduced for an ellipse and here is used for our case as

$$\zeta = 0.5l(z + z^{-1}), \ \ {}^{z}_{z^{-1}} = \left(\zeta \pm \sqrt{\zeta^2 - l^2}\right)/l \tag{3.7}$$

we put the second of these expressions into (3.6) and using (3.5) we receive

$$w(\zeta) = Q \ln\left(\zeta/l + \sqrt{(\zeta/l)^2 - 1}\right)/\pi + Gu_o,\ \tau_\xi - i\tau_\eta = Q/\pi\sqrt{\zeta^2 - l^2} = w'(\zeta). \tag{3.8}$$

Along axis ξ we compute as follows

$$\begin{aligned}&\text{At}/\xi/ < l : u_z = u_o,\ \tau_\xi = 0,\ \tau_\eta = Q/\pi\sqrt{l^2 - \xi^2},\\ &\text{At}/\xi/ > l : u_z = u_o - (Q/G\pi)\ln(\xi/l + \sqrt{(\xi/l)^2 - 1}), \tau_\eta = 0, \tau_\xi = Q/\pi\sqrt{\xi^2 - l^2}.\end{aligned} \tag{3.9}$$

In the same manner the displacement and stresses in any part of the massif can be found. The most dangerous points are $\eta = 0$, $/\xi/ = l$ and in order to investigate the fracture process there it is convenient to use decomposition $\xi - l = re^{i\theta}$ which gives according to expressions (3.5), (3.8) and (2.52)

$$u_z = u_o - (Q/G\pi)\sqrt{2r/l}\cos(\theta/2),\ \begin{matrix}\tau_r\\ \tau_\theta\end{matrix} = Q/\pi\sqrt{2rl} \times \begin{matrix}-\cos(\theta/2)\\ \sin(\theta/2)\end{matrix},\ \tau_e = Q/\pi\sqrt{2rl}.$$

From the third of these relations we can see that the condition τ_e = constant gives a circumference with a center at $/\xi/ = l$ where a fracture or plastic strains should begin. Now following the Griffith's idea (see Sect. 1.5.2) we find critical load similar (1.23) (see also Sect. 3.1.4)

$$Q_* = \sqrt{2G\gamma_s \pi l} \tag{3.10}$$

3.1.3 Deformation of Massif with Circular Hole of Unit Radius

In this case (Fig. 3.3) the boundary conditions are:

$$\tau_\eta|_{\zeta=\infty} = \tau_\zeta,\ \tau_\rho|_{\rho=1} = 0. \tag{3.11}$$

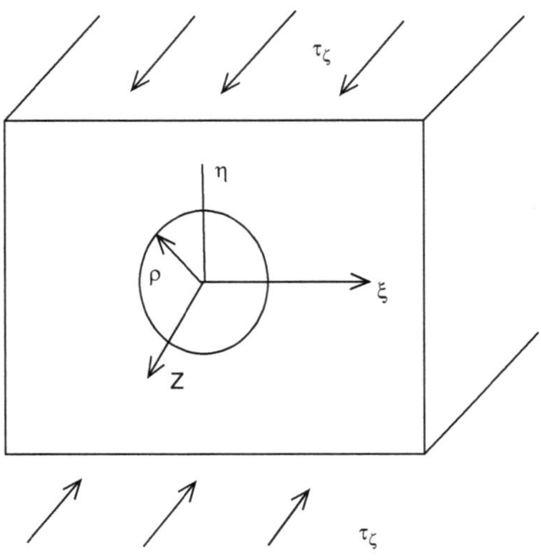

Fig. 3.3. Massif with circular hole

We seek the solution in a form

$$w'(\zeta) = \sum_{n=0}^{\infty} A_n/\zeta_n; \tag{3.12}$$

and the first condition (3.11) gives immediately $A_0 = -i\tau_\zeta$. Now with the help of (3.5) we rewrite the second (3.11) as $\mathrm{Re}(e^{i\theta}w'(e^{i\theta})) = 0$ from that we have $A_2 = -i\tau_\zeta$ and all other factors are equal to zero. So, we receive

$$w'(\zeta) = -i\tau_\zeta(1+\zeta^{-2}),\ w(\zeta) = -i\tau_\zeta(\zeta - \zeta^{-1}). \tag{3.13}$$

For example at $\eta = 0$ we compute by (3.5)

$$\tau_\xi = u_z = 0,\ \tau_\eta = \tau_\zeta(1+\xi^{-2})$$

and $\tau_\eta(1) = 2\tau_\zeta$ – the dangerous point.

3.1.4 Brittle Rupture of Body with Crack

This problem is very significant in the Mechanics of Fracture. In the literature it is usually named as the third task of cracks. Its solution can be received by conformal transformation of the second relation (3.13) with the help of Zhoukovski's expression (3.7) in which the variables ζ and z are interchanged. So, with consideration of condition $\tau_o = 2\tau_\zeta/l$ we have (Fig. 3.4)

$$w'(z) = -\tau_o z/\sqrt{l^2 - z^2},\ w(z) = \pm\,\tau_o\sqrt{l^2 - z^2}. \tag{3.14}$$

It is not difficult to prove that the first expression (3.14) can be got from the second (3.8) after replacing in it ζ, l, Q/π by $1/z$, $1/l$, τ_o/l respectively.

At $x = 0$ we determine from (3.14)

$$u_z = \pm\,(\tau_o/G)\sqrt{l^2 + y^2},\ \tau_y = \tau_o y/\sqrt{l^2 + y^2},\ \tau_x = 0. \tag{3.15}$$

Along the other axis (x) we find similarly

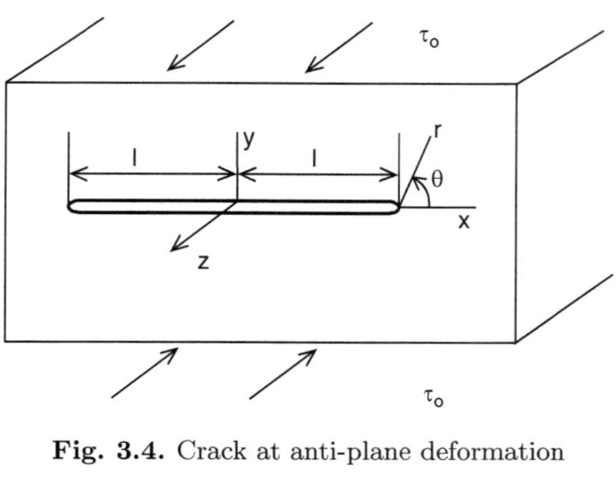

Fig. 3.4. Crack at anti-plane deformation

$$u_z = \pm(\tau_o/G)\sqrt{l^2 - x^2},\ \tau_y = 0,\ \tau_x = -\tau_o x/\sqrt{l^2 - x^2}(/x/ < l),$$
$$u_z = 0,\ \tau_x = 0,\ \tau_y = \tau_o x/\sqrt{x^2 - l^2}(/x/ > l). \tag{3.16}$$

According to the Clapeyron's theorem we can compute the work which is done by stress τ_y at its decrease from τ_o to zero which corresponds to a formation of the crack as

$$W = \tau_o \int_{-l}^{l} u_z dx = \pi(\tau_o)^2 l^2/2G. \tag{3.17}$$

When the crack begins to propagate an increment of the work becomes equal to a stretching energy $4\gamma_s dl$ where γ_s is this energy per unit length. From this condition we find critical stress

$$\tau* = 2\sqrt{\gamma_s G/\pi l}. \tag{3.18}$$

From the strength point of view stresses and strains in the edge of the crack are of the greatest interest. To find them we use the asymptotic approach as in Sect. 3.1.2 that in polar coordinates r, θ (see Fig. 3.4) according to expressions (3.5), (3.14) gives

$$u_z = \tau_o\sqrt{2rl}\sin(\theta/2)/G,\ \tau_r = \sqrt{l/2r\tau_o}\sin(\theta/2),$$
$$\tau_\theta = \sqrt{l/2r\tau_o}\cos(\theta/2),\ \tau_e = \tau_o\sqrt{l/2r}. \tag{3.19}$$

From the fourth of these relations we can see that in this case condition τ_e = constant represents also a circumference with the center in the top of crack.

Since the largest part of the energy concentrates near the crack edges we can use expressions (3.19) for the computation of $\tau*$. When the crack grows we should put in the first relation (3.19) $\theta = \pi$, $r = dl - x$ and in the third one – $\theta = 0$, $r = x$, then the increment of the work at the crack propagation is equal to that in (3.17) as follows

$$dW = \int_0^{dl} \tau_\theta(0)u_z(\pi)dx = ((\tau_o)^2/G)ldl \int_0^1 \sqrt{(1-\xi)/\xi}d\xi$$

or after integration

$$dW = (\tau_o^2/2G)\pi ldl. \tag{3.20}$$

Now we introduce an intensity factor $K_3 = \tau_o\sqrt{\pi l}$ and equaling dW to the stretching energy $2\gamma_s dl$ we find $K_{3*} = 2(\gamma_s G)^{0.5}$ after that the strength condition may be written in form

$$K_3 \leq K_{3*}$$

where factor K_{3*} is determined by the properties of the continuum and its value can be found experimentally in elastic or plastic range /17/. The tests show that the condition K_{3*} = constant fulfils well enough for brittle bodies only. However (3.20) characterizes a resistance of the material to the crack propagation (the so-called fracture toughness).

3.1.5 Conclusion

Problems of anti-plane deformation are ones of the simplest in the Mechanics of Continuum and Fracture. But their solutions have practical and theoretical value. Many processes in the earth (a loss of structures stability, landslides etc.) occur due to shear stresses. Later on we will consider the problems above in a non-linear range and the analogy between the punch movement and crack propagation will be used for finding the solution of one of them when a result of the other is known. Moreover there is a similarity between these results and ones in the plane deformation will be also of great importance.

3.2 Plane Problem

3.2.1 Wedge Under One-Sided Load

In this case we suppose that stresses and strains do not depend on coordinate r (Fig. 3.5) and expressions (2.67), (2.71) become

$$d\tau/d\theta + \sigma_r - \sigma_\theta = 0,\ d\sigma_\theta/d\theta + 2\tau = 0,\ d\varepsilon_r/d\theta = \gamma - C/G \tag{3.21}$$

where C is a constant. Combining these relations with the Hooke's law (2.62) (in polar coordinates) we receive equation

$$d^2\tau/d\theta^2 + 4\tau = 4C$$

which has the solution with consideration of boundary condition $\tau(\pm\lambda) = 0$ in form

$$\tau = C_o(\cos 2\theta - \cos 2\lambda) \tag{3.22}$$

and from (3.21)

$$\begin{matrix}\sigma_r \\ \sigma_\theta\end{matrix} = C_1 + C_o(2\theta\cos 2\lambda \pm \sin 2\theta). \tag{3.23}$$

Finally boundary conditions $\sigma_\theta(-\lambda) = 0$, $\sigma_\theta(\lambda) = -p$ give the values of constants as

$$C_o = 0.5p(\sin 2\lambda - 2\lambda\cos 2\lambda)^{-1},\ C_1 = -p/2 \tag{3.24}$$

and according to (2.65)

$$\tau_e = 0.5p\sqrt{(1 - 2\cos 2\theta\cos 2\lambda + \cos^2 2\lambda)}/(\sin 2\lambda - 2\lambda\cos 2\lambda). \tag{3.25}$$

The analysis of (3.25) shows that at $\lambda = \pi/4$ the maximum shearing stress is the same in the whole wedge and it is equal to p/2. At other $\lambda > \pi/4$ τ_e reaches his maximum only on axis $\theta = 0$ (interrupted by points line in Fig. 3.5).

In order to compute displacements we firstly determine the strains according to the Hooke's law (2.62) as

$$\begin{gathered}\begin{matrix}\varepsilon_r \\ \varepsilon_\theta\end{matrix} = (-0.5p(1-2\nu) + C_o(2(1-2\nu)\theta\cos 2\lambda \pm \sin 2\theta)/2G), \\ \gamma = C_o(\cos 2\theta - \cos 2\lambda)/G.\end{gathered} \tag{3.26}$$

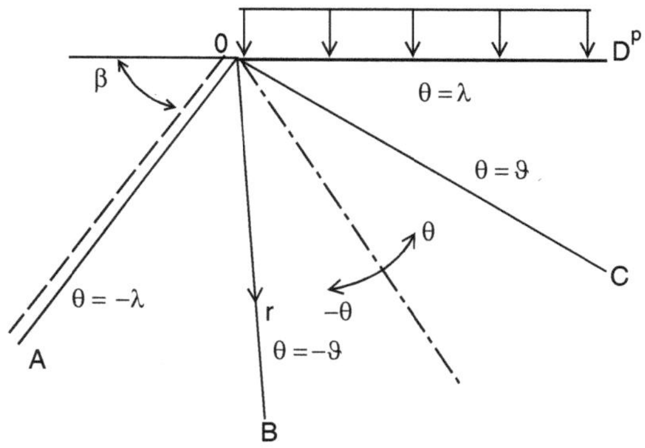

Fig. 3.5. Wedge under one-sided load

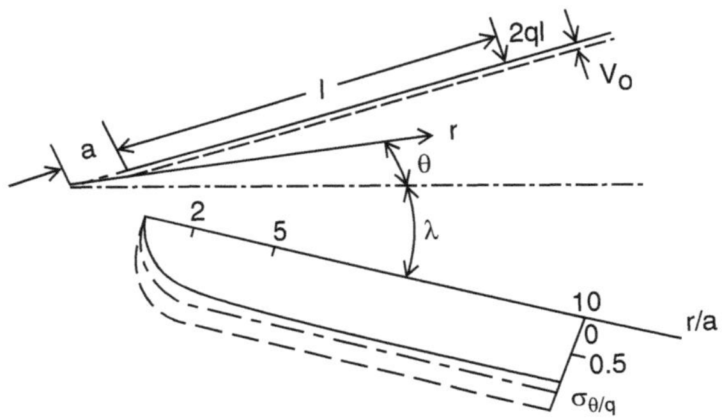

Fig. 3.6. Wedge pressed by inclined plates

Using (2.69), neglecting the constant displacement and excluding infinite values at $r = 0$ we receive

$$\begin{aligned} u_r &= r(-0.5p(1-2\nu) + C_o(2(1-2\nu)\theta\cos 2\lambda + \sin 2\theta))/2G, \\ u_\theta &= C_o r(\cos 2\theta - 4(1-\nu)\ln(r/d)\cos 2\lambda)/2G. \end{aligned} \tag{3.27}$$

where d is the the value of r where points have no tangential displacements for incompressible material ($\nu = 0.5$) relations (3.26), (3.27) become much simpler.

3.2.2 Wedge Pressed by Inclined Plates

General Case

Let plates move parallel to their initial position (broken straight lines in Fig. 3.6) with displacement $V(\lambda) = V_o$. Then according to (2.70), (2.71) at $u_\theta = V(\theta)$ and (2.69) we receive

$$u_r = U(\theta)/r - V', \ \varepsilon_r = -\varepsilon_\theta = -U/r^2, \ \gamma = U'/r^2 - f(\theta)/r \tag{3.28}$$

where $f = V + V''$ and from (2.66) at $m = 1$, $\Omega = 1/3G$ we have at $\tau \equiv \tau_{r\theta}$

$$\tau = G(U'/r^2 - f/r),\ \sigma_r - \sigma_\theta = -4GU/r^2 \qquad (3.29)$$

that gives together with (2.67)

$$\sigma_\theta = F(r) + (G/r)\int f d\theta,\ \sigma_r = F(r) + (G/r)\int f d\theta. \qquad (3.30)$$

Putting stresses according to (3.29), (3.30) into the first static law (2.67) we receive an equality

$$r^3 dF/dr - Gr\left(f' + \int f d\theta\right) = -G(U'' + 4U)$$

both parts of which must be equal to the same constant, e.g. n and

$$f' + \int f d\theta = 0.$$

With the consideration of anti-symmetry and symmetry conditions we determine

$$f = -C\sin\theta,\ U = -D\cos 2\theta - n/4G$$

where C, D are constants. In order to find n we use a stick demand /18/ $U(\lambda) = 0$ which gives

$$U = -D(\cos 2\theta - \cos 2\lambda),$$

and, consequently, – the stresses as

$$\begin{aligned}
\tau &= G((2D/r^2)\sin 2\theta + (C/r)\sin\theta),\\
\sigma_\theta &= A + 2(GD/r^2)\cos 2\lambda + (GC/r)\cos\theta,\\
\sigma_r &= A - 2(GD/r^2)(\cos 2\lambda - 2\cos 2\theta) + (GC/r)\cos\theta
\end{aligned} \qquad (3.31)$$

where constants A, C, D should be determined from condition $\sigma_\theta(a, \lambda) = 0$ and integral static equations

$$\begin{aligned}
&\int_{-\lambda}^{\lambda} \sigma_r(a, \theta)\cos\theta d\theta = 0,\\
&\int_{a}^{a+l} \sigma_\theta(r, \lambda) dr = -ql.
\end{aligned} \qquad (3.32)$$

Putting in (3.32) stresses according to (3.31) we derive at $\theta = \lambda$

$$\sigma_\theta = -(q/B_o)(\Lambda(1 - a^2/r^2)\cos 2\lambda + (1 - a/r)\cos\lambda) \qquad (3.33)$$

where

$$B_o = \Lambda(1 + a/l)^{-1}\cos 2\lambda + (1 - (a/l)\ln(l/a + 1)\cos\lambda) \qquad (3.34)$$

and

$$\Lambda = 3(\cos\lambda - \lambda/\sin\lambda)/16\sin 2\lambda.$$

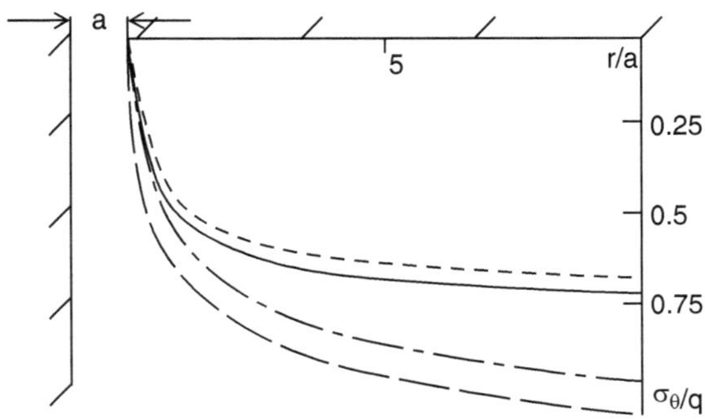

Fig. 3.7. Model of retaining wall

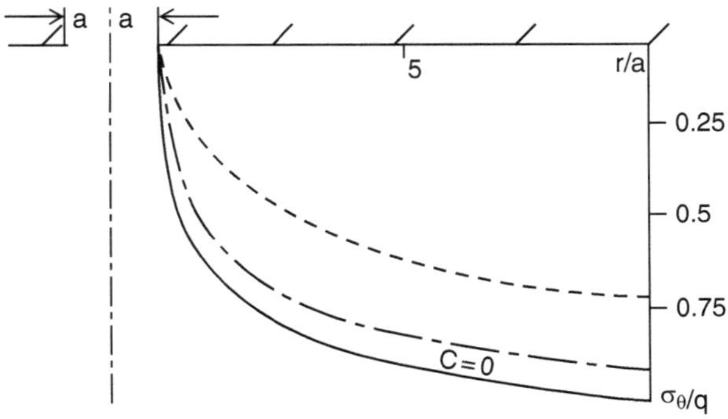

Fig. 3.8. Model of two foundations

In practice $l \gg a$ and (3.34) becomes

$$B_o = \Lambda \cos 2\lambda + \cos \lambda \tag{3.35}$$

The diagrams $\sigma_\theta(r/a)$ at $l/a = 9$ are given by solid lines in Figs. 3.6–3.8 for $\lambda = \pi/6$, $\lambda = \pi/4$ (a model of a retaining wall) and $\lambda = \pi/2$ (a flow of the material between two foundations) respectively.

From (3.31), (2.65) with consideration of D-value we compute maximum shearing stress as

$$\tau_e = (q/B_o)((\Lambda^2 \cos 2\lambda - \cos 2\theta)^2 + (\Lambda \sin 2\theta + \sin \theta)^2)^{0.5.} \tag{3.36}$$

To find maximum of τ_e we use condition $d\tau_e/d\theta = 0$ which gives $\theta = 0$ and equation

$$\cos 2\theta + (1/6)(4\Lambda \cos 2\lambda + 1/\Lambda)\cos\theta - 1/3 = 0. \tag{3.37}$$

Investigations show that at $\lambda < 5\pi/12$ an impossible condition $\cos\theta > 1$ takes place and hence τ_e should be found from (3.36) at $\theta = 0$ as follows

$$/\tau_e/ = 2q\Lambda(\sin^2 \lambda)/B_o.$$

But at $\lambda > 5\pi/12$ $\max \tau_e$ is determined by (3.36) with θ from (3.37). Diagram $\max \tau_e(\lambda)$ is given in Fig. 3.9 by solid line 1.

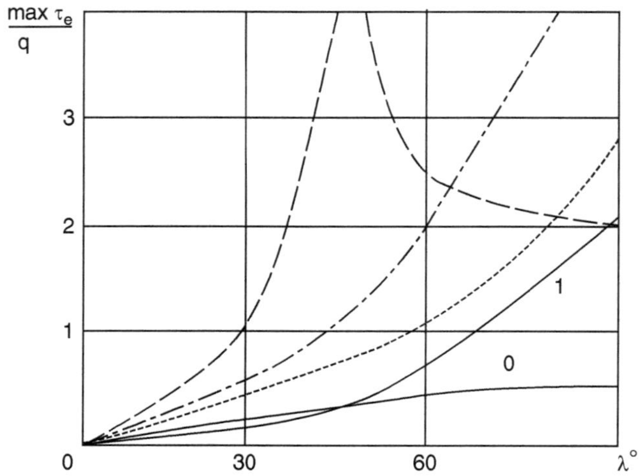

Fig. 3.9. Diagram $\max \tau_e(\lambda)$

Some Particular Cases

Besides this common solution it is interesting to study two simpler options: $C = 0$ (a compulsory flow of the material between immovable plates) and $D = 0$ (when plates move and the compulsory flow is negligible). In the both cases we use expressions for stresses (3.31), condition $\sigma_\theta(a, \lambda) = 0$ and the second static equation (3.32). For the first option we have

$$\sigma_\theta = -q(1 - a^2/r^2)/B_1, \tag{3.38}$$

$$\tau_e = (q/B_1)(1 - 2\cos 2\theta/\cos 2\lambda + 1/\cos^2 2\lambda)^{0.5} \tag{3.39}$$

where

$$B_1 = 1 + a/l.$$

At $l >> a$ $B_1 = 1.163$ and diagrams $\sigma_\theta(r/a)$ are shown in Figs. 3.6 and 3.7 by broken lines. Condition $d\tau_e/d\theta = 0$ gives demand $\sin 2\theta = 0$ with consequent solutions $\theta = 0$ and $\theta = \pi/2$ but calculations show that only the first of them gives to τ_e the maximum value which is

$$\max \tau_e = (q/B_1)(1 - 1/\cos 2\lambda)$$

Diagram $\max \tau_e(\lambda)$ according to this relation is drawn in Fig. 3.9 by broken line and we can see that at $\lambda = \pi/4$ it tends to infinity.

For the case $D = 0$ we derive in a similar way

$$\sigma_\theta = -(q/B_2)(1 - (a\cos\theta)/(r\cos\lambda)), \tag{3.40}$$

$$\tau_e = (q/B_2)(\sin\theta/\cos\lambda) \tag{3.41}$$

where

$$B_2 = 1 - (a/l)\ln(1 + l/a)$$

The highest value of σ_θ is at $\theta = \lambda$ and since at $l \gg a$ $B_2 = 1$. Diagrams $\sigma_\theta(r/a)$ are given by interrupted by points lines in Figs. 3.6–3.8. Maximum τ_e takes place at $\theta = \lambda$ and the consequent diagram is shown in Fig. 3.9 by interrupted by points curve which tends to infinity at $\lambda = \pi/2$. So we can conclude that $\max \tau_e$—values can be high and plastic deformations are expected in some zones.

Case of Parallel Plates

As an interesting particular case we consider a version of parallel plates (Fig. 3.10). We take u_y as a function of y only and according to incompressibility equation $\varepsilon_x + \varepsilon_y = 0$ as well as symmetry and stick (at $y = h$) conditions we find

$$u_x = 3V_o x(h^2 - y^2)/2h^3,\ u_y = V_o y(y^2 - 3h^2)/2h^3 \tag{3.42}$$

where V_o is a velocity of plates movement. Then we compute strain by relations (2.60) and stresses by the Hooke's law (2.62) and static equations as follows

$$\begin{aligned} \tau_{xy} &= -3V_o Gxy/h^3, \\ \sigma_x &= 3GV_o(3(h^2 - y^2) + x^2 - l^2)/2h^3, \\ \sigma_y &= 3GV_o(y^2 - h^2 + x^2 - l^2)/2h^3. \end{aligned} \tag{3.43}$$

Here condition $\sigma_y(l, h) = 0$ is used. Integral equilibrium law

$$\int_{-l}^{l} \sigma_y(x, h)dx = -P \tag{3.44}$$

gives

$$P = 2GV_o l^3/h^3. \tag{3.45}$$

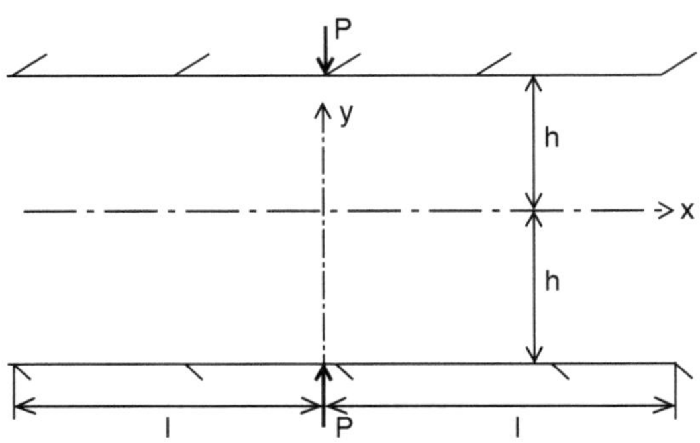

Fig. 3.10. Compression of layer by parallel plates

3.2.3 Wedge Under Concentrated Force in its Apex: Some Generalizations

General Case

In this problem we have boundary conditions: $\sigma_\theta = \tau_{r\theta} = 0$ at $\theta = \pm\lambda$. Since angle λ is arbitrary we can suppose that σ_θ and $\tau_{r\theta}$ are absent at any θ. That allows to seek function Φ from (2.75) in form

$$\Phi = \mathrm{r}f(\theta) + C \tag{3.46}$$

where C is a constant. Now we put (3.46) into (2.74) which gives $f'' + f = 0$ and with consideration of symmetry as well as static condition at $\alpha_o = 0$ (Fig. 3.11) we receive

$$\sigma_r = -P(\cos\theta)/r(\lambda + 0.5\sin 2\lambda). \tag{3.47}$$

The strains and displacements can be determined as usual.
The particular case $\lambda = \pi/2$ is of great interest

$$\begin{aligned}\sigma_r &= -2P(\cos\theta)/\pi r,\\ \varepsilon_r &= -2P(\cos\theta)/E\pi r,\\ \varepsilon_\theta &= 2P\nu(\cos\theta)/E\pi r.\end{aligned} \tag{3.48}$$

Using relations (2.69) we find displacements u_r, u_θ /5/ which we write for the edge of the semi-plane as follows

$$u_x \equiv u_r = -P(1-\nu)/2E\pi,\ u_y \equiv u_\theta = (2P/\pi E)\ln(x/d) \tag{3.49}$$

where d is the coordinate on axis y which has no vertical movement.

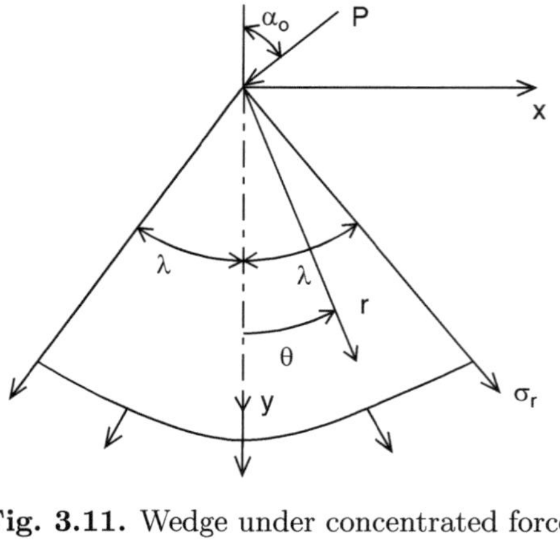

Fig. 3.11. Wedge under concentrated force

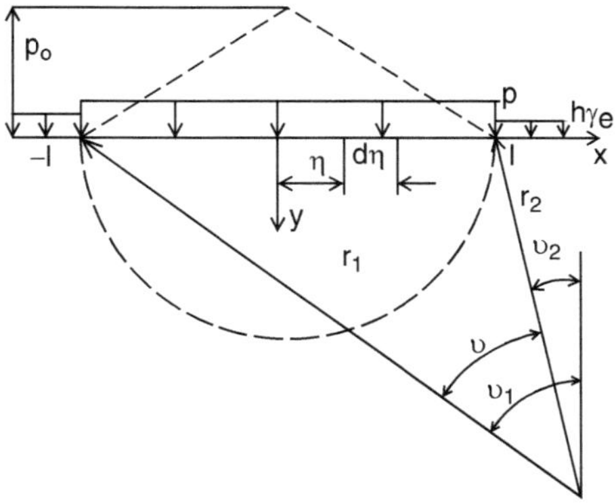

Fig. 3.12. Deformation under flexible load

Case of Distributed Load

If the load is distributed in interval (a, b) we replace in the relations above P by p$d\eta$ and integrate as follows

$$u_y = u_o + (2/\pi E) \int_a^b p(\eta) \ln(x - \eta) d\eta \tag{3.50}$$

and for the case p = constant at $-l \leq \eta \leq l$ (Fig. 3.12) we compute

$$u_y = u_o - 2(pl/\pi E)((x/l+1)\ln(x/l+1) + (1-x/l)\ln(x/l-1))(/l/ \leq x). \tag{3.51}$$

In the same manner stresses under a flexible load in an interval $(-l, l)$ can be found. We begin with the first relation (3.48) and according to (2.72) we write

$$\sigma_y = -(2P/\pi r)\cos^3\theta = -2Py^3/\pi r^4,$$
$$\tau_{xy} = -2Pxy^2/\pi r^4,\ \sigma_x = -2Px^2y/\pi r^4.$$

Now as before we replace P by p$d\eta$ and summarize the loads as follows:

$$\sigma_y = -\left(2py^3/\pi\right) \int_{-1}^{1} \left(y^2 + (x-\eta)^2\right)^{-2} d\eta$$

or after integration:

$$\sigma_y = -p(\upsilon_1 - \upsilon_2 + 0.5(\sin 2\upsilon_1 - \sin 2\upsilon_2))/\pi. \tag{3.52}$$

Similarly we find:

$$\sigma_x = -p(\upsilon_1 - \upsilon_2 + 0.5(\sin 2\upsilon_2 - \sin 2\upsilon_1))/\pi,\ \tau_{xy} = p(\sin^2\upsilon_1 - \sin^2\upsilon_2)/\pi. \tag{3.53}$$

According to (2.65) and (2.72) we compute maximum shearing stress and main components as:

$$\tau_e = p \sin(\upsilon_1 - \upsilon_2)/\pi,$$
$$\begin{matrix}\sigma_1 \\ \sigma_3\end{matrix} = \frac{p(\upsilon_2 - \upsilon_1 \pm \sin(\upsilon_2 - \upsilon_1))}{\pi}. \quad (3.54)$$

The biggest τ_e is p/π and it realizes on the curve $x^2 + y^2 = l^2$ (broken line in Fig. 3.12). Hence if strength condition $\tau_e = \tau_{yi}$ is used a sliding along this circumference should be considered. This result was particularly applied to an appreciation of the earth resistance to mountains movement.

The First Ultimate Load

For the foundation with width 2l and depth h we must add load $\gamma_e h$ outside the main one (p in Fig. 3.12). We suppose the hydrostatic distribution of the soil's weight $\sigma_{3s} = \sigma_{1s} = \gamma_e(h + y)$, We take also compressive stresses as positive and rewrite the second relation (3.54) in the following way:

$$\begin{matrix}\sigma_1 \\ \sigma_3\end{matrix} = \frac{(p - \gamma_e h)(\upsilon \pm \sin \upsilon)}{\pi} - \gamma_e(h + y) \quad (3.55)$$

where $\upsilon = \upsilon_1 - \upsilon_2$. Now we put these stresses in the ultimate equilibrium condition (see broken line in Fig. 1.22) in form:

$$\sigma_1 - \sigma_3 = 2(\sigma_m + c \cot \varphi) \sin \varphi. \quad (3.56)$$

That gives expression:

$$((p - \gamma_e h)/\pi) \sin \upsilon - (((p - \gamma_e h)/\pi)\upsilon + \gamma_e(h + y)) \sin \varphi = c \cos \varphi$$

from which we can find equation of the boundary curve where the first plastic strains can appear:

$$y = ((p - \gamma_e h)/\pi\gamma_e)(\sin \upsilon / \sin \varphi - \upsilon) - c/\gamma_e \tan \varphi - h. \quad (3.57)$$

Now we use the condition $dy/d\upsilon = 0$ which gives $\cos \upsilon = \sin \varphi$ or:

$$\upsilon = \pi/2 - \varphi \quad (3.58)$$

Putting (3.58) into (3.57) we derive:

$$y_{max} = ((p - \gamma_e h)/\pi\gamma_e)(\cot \varphi + \varphi - \pi/2) - c(\cot \varphi)/\gamma_e - h.$$

If we decide to find a load at which plastic deformation does not begin in any point we must suppose $y_{max} = 0$ that leads to minimum load:

$$\min p_{yi} = \pi(\gamma_e h + c \cot \varphi)/(\cot \varphi + \varphi - \pi/2) + \gamma_e h$$

which is always called in the literature as the first critical load that was found by Professor Pousyrevski in 1929. At h = 0 it becomes:

$$\min p_{yi} = \pi c/(1 + (\varphi - \pi/2)\tan\varphi) \tag{3.59}$$

and this relation can not be applied to soils without cohesiveness. If $\varphi = 0$ it coincides with value p_u (ultimate load) following from (3.54). At other φ factor c may be represented as $a_1\tau_{yi}$ where a_1-values are given in Table here. If a_1 is less $p_{yi} < p_u$.

φ	0	5	10	15	20	25	30	35	40	45
a_1	1	0.87	0.75	0.64	0.56	0.47	0.40	0.32	0.27	0.215
a_2	0.61	0.49	0.38	0.29	0.21	0.15	0.10	0.07	0.04	0.024
a_1/a_2	1.64	1.80	2.0	2.27	2.62	3.11	3.79	4.83	6.41	9.174

To take more real approach we suppose that:

$$p(\eta) = p_o(1 - \eta/l)$$

(broken straight lines in Fig. 3.12). Computations as before give:

$$\begin{aligned}
\sigma_y &= p_o((1-x)(\upsilon_2 - \upsilon_1 + 0.5(\sin 2\upsilon_1 - \sin 2\upsilon_2)) + y(\cos^2\upsilon_1 - \cos^2\upsilon_2))/\pi l,\\
\sigma_x &= p_o((1-x)(\upsilon_2 - \upsilon_1 - 0.5(\sin 2\upsilon_1 - \sin 2\upsilon_2))\\
&\quad + y(\cos^2\upsilon_2 - \cos^2\upsilon_1 + \ln(r_2/r_1)))/\pi l,\\
\tau_{xy} &= p_o((1-x)(\cos^2\upsilon_2 - \cos^2\upsilon_1) + y(\upsilon_2 - \upsilon_1 + 0.5(\sin 2\upsilon_2 - \sin 2\upsilon_1)))/\pi l.
\end{aligned}$$

Expression (2.65) for τ_e is somewhat complex. To simplify the problem we find it on axis y (at x = 0) which gives expression at y/l = u

$$\tau_e = (p_{oy}/\pi)\sqrt{1(u + 1/u)^2 + (u\tan^{-1}(1/u) - 1(u + 1/u)^2}$$

that grows with an increase of y and reaches its biggest value in infinity. So according to this approach the mountain range is stable.

3.2.4 Beams on Elastic Foundation

We model such a foundation as a set of closely spaced separate springs (Fig. 3.13) which exert reactions r(x) = kv(x) where k is a modulus of an elastic bed and v – a deflection of a beam. If distributed load is p(x) total forces acting on the beam are q = p − r and according to the Strength of Material law $q = -M''$ where M is linked with the deflection by expression (1.54) at m = 1 as

$$v'' = -M/EI. \tag{3.60}$$

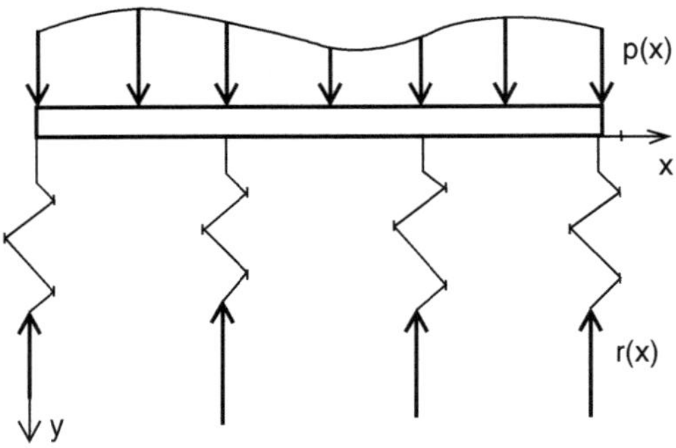

Fig. 3.13. Beam on elastic foundation

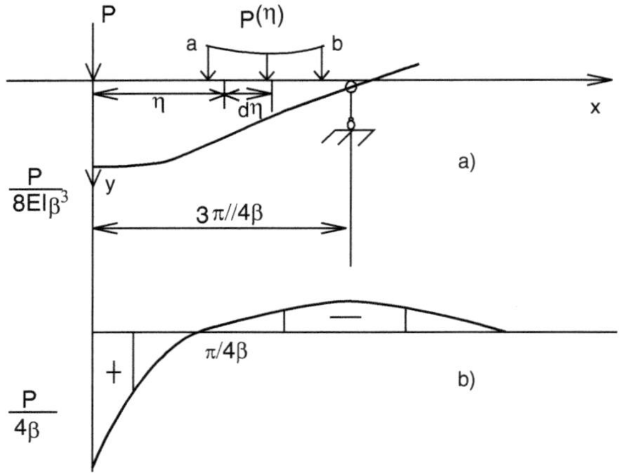

Fig. 3.14. Concentrated load in origin

Here E-elasticity modulus of the beam material. Combining all these relations we receive linear differential equation of the fourth order for the beam on elastic foundation as follows:

$$v^{IV} + 4\beta^4 v = p(x)/EI$$

where the propotionality factor is:

$$4\beta^4 = k/EI \tag{3.61}$$

The general solution of (3.61) is:

$$v = Ae^{\beta x}\cos\beta x + Be^{\beta x}\sin\beta x + Ce^{-\beta x}\cos\beta x + De^{-\beta x}\sin\beta x + v_p. \tag{3.62}$$

Here A, B, C, D – constants and v_p is a particular solution.

Now we consider the important example of concentrated load P in the origin of the coordinate system x, y (Fig. 3.14, a). In this case $p(x) = 0$ and hence $v_p = 0$. Since displacement in infinity has finite value we must suppose

A = B = 0. Constants C, D can be found from conditions $v'(0) = 0$ and $v'''(0) = -Q/EI = -P/2EI$ where Q is a shearing force (see Sect. 1.5.2.). As a result we have the following solution for the right part:

$$v = Pe^{-\beta x}(\cos\beta x + \sin\beta x)/8\beta^3 EI \tag{3.63}$$

(broken line in Fig. 3.14, a) with maximum deflection at $x = 0$ as:

$$v_{max} = P/8\beta^3 EI \tag{3.64}$$

and $v = 0$ in point $x = 3\pi/4\beta$.

Using law (3.60) we find from relation (3.63) expression for bending moment (Fig. 3.14, b)

$$M = Pe^{-\beta x}(\cos\beta x - \sin\beta x)/4\beta \tag{3.65}$$

with maximum value $M_{max} = P/4\beta$ and $M = 0$ in points $x = \pi/4\beta$, $x = 5\pi/4\beta$ etc. (Fig. 3.14, b).

In order to appreciate the role of the elastic foundation we compare this result to a similar bending solution for a beam with supports at $x = \pm 3\pi/4\beta$ (the moments there are equal to $-2^{0.5}Pe^{-3\pi/4}/4\beta = -0.007Pl$ and they can be neglected). According to well-known relations of the Strength of Materials for such a beam without elastic foundation $M_{max} = Pl/4$ and $v_{max} = Pl^3/48EI$. For our task $\beta = 3\pi/2l$ and we conclude that the elastic foundation decreases the maximum moment and deflection in 5.7, 17.4 times. So the role of the bed is significant.

Similarly to the previous section we can consider a distributed load in (3.63) as

$$v(x) = (1/8\beta^3 EI)\int_a^b e^{-\beta(x-\eta)}p(\eta)(\cos\beta(x-\eta) + \sin\beta(x-\eta))d\eta,$$

When $p =$ constant at $-l < \eta < l$ we compute

$$v(x) = pe^{-\beta x}(e^{\beta l}\cos\beta(x-l) - \cos\beta x)/2k$$

Condition $v'(0) = 0$ gives for the right side of the beam equation

$$\cos\beta l - \sin\beta l = e^{\beta l}/2\beta l. \tag{3.66}$$

Combining these two relations and using (3.61) we find expressions similar (3.63), (3.65) as

$$v = e^{\beta l}p(\sin\beta l)e^{-\beta x}(\sin\beta x + \cos\beta x)/2k,$$
$$M = pe^{\beta l}(\sin\beta l)e^{-\beta x}(\cos\beta x - \sin\beta x)/4\beta^2.$$

To compare these results with the previous case we put here $p = P/2l$ and apply (3.61). We receive respectively (3.63), (3.65) with multiplier

$$(e^{\beta l}/2\beta l)\sin\beta l.$$

Solutions of (3.66) are 0 and near $5\pi/4$, $9\pi/4\ldots$ but only the third one can be used when the multiplier is approximately 60 and so the length of the beam must be taken in consideration.

3.2.5 Use of Complex Variables

General Expressions

Main equations of the plane problem are given by relations (2.59)–(2.62). Similarly to the case of a longitudinal shear it is convenient to use here complex variables when law (2.63) becomes

$$\partial^4\Phi/\partial z^2\partial\bar{z}^2 = 0 \tag{3.67}$$

where the line over a value means that we must change in it i by −i. The solution of (3.67) is obvious, containing two functions, e.g. F(z) and χ(z). When they are known stresses and displacements can be found according to relations

$$\sigma_x + \sigma_y = 4\mathrm{Re}F'(z),$$
$$\sigma_y - \sigma_x + 2i\tau_{xy} = 2(\bar{z}F''(z) + \chi'(z)), \tag{3.68}$$
$$2G(u_x + iu_y) = \kappa F(z) - z\overline{F'(z)} - \overline{\chi(z)}. \tag{3.69}$$

Here κ is equal to 3−4ν and (3 − ν)/(1 + ν) for plane deformation and the same stress state respectively.

Functions F(z) and χ(z) must be found from border demands that for the first boundary problem (see sub-chapter 2.4) is

$$\sigma_y + i\tau_{xy} = F'(z) + zF''(z) + F'(z) + \chi'(z) \tag{3.70}$$

or as a resultant of forces on an arc ab:

$$P_x + iP_y = -(F(z) + z\overline{F'(z)} + \overline{\chi(z)})|_a^b. \tag{3.71}$$

Tension of Plate with Circular Hole

As an example we study a plane with a circular hole of a radius R (Fig. 3.15) under tension by stresses σ in infinity (the problem was solved by G. Kirsch in 1898). According to border condition (3.71) we have at $z = Re^{i\theta}$

$$F(z) + z\overline{F'(z)} + \overline{\chi(z)} = 0. \tag{3.72}$$

Similar to the case of massif with circular hole at anti-plane deformation we seek the solution in series as:

$$F(z) = Az + F_o(z), \chi(z) = Bz + \chi_o(z)$$

where A, B – constants and

$$F_o(z) = \sum_{n=0}^{\infty} A_n/z^n, \chi_o(z) = \sum_{n=0}^{\infty} B_n/z^n. \tag{3.73}$$

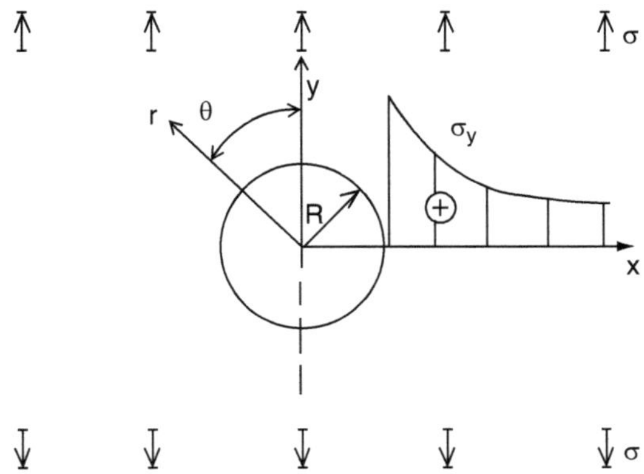

Fig. 3.15. Tension of plate with circular hole

From expressions (3.68) we find at $z \to \infty A = \sigma/4$, $B = \sigma/2$. Then we put $F_o(z)$, $\chi_o(z)$ into (3.72) that gives:

$$\sigma z/4 + F_o(z) + \sigma z/4 + z\overline{F'_o(z)} + \sigma \bar{z}/2 + \overline{\chi_o(z)} = 0. \tag{3.74}$$

Now we consider in (3.74) condition $z = R^2/\bar{z}$ and compute

$$F_o(z) + R^2\overline{F'_o(z)}/\bar{z} + \overline{\chi_o(z)} = -\sigma R^2/2z - \sigma R^2/2\bar{z}. \tag{3.75}$$

Taking into account A, B – values above and integrating the derivative $F'_o(z) = \sigma R^2/2z^2$ we finally receive:

$$F(z) = \sigma z/4 - \sigma R^2/2z, \chi(z) = -\sigma R^2/2z + \sigma z/2 - \sigma R^4/2z^3. \tag{3.76}$$

From (3.69) we have:

$$2G(u_x + iu_y) = (\kappa - 1)\sigma z/4 - \kappa\sigma R^2/2z - \sigma z R^2/2\bar{z}^2 - \sigma z R^2/2\bar{z}^2 - \sigma\bar{z}/2 \\ + \sigma R^2/2\bar{z} + \sigma R^4/2\bar{z}^3. \tag{3.77}$$

To determine the stresses we differentiate functions $F'(z)$, $\chi(z)$ as:

$$F''(z) = -\sigma R^2/z^3, \chi'(z) = \sigma/2 + \sigma R^2/2z^2 + 3\sigma R^4/2z^4. \tag{3.78}$$

Then relations (3.68)–(3.70) can be used to determine stresses and displacements. For example at $y = 0$ we have:

$$\sigma_y = \sigma(1 + R^2/2x^2 + 3R^4/2x^4) \tag{3.79}$$

(shaded diagram in Fig. 3.15) with maximum σ_y at $x = R$ as:

$$\max \sigma_y = 3\sigma. \tag{3.80}$$

If $x = 0$, $y = R$ we find in a similar manner $\tau_{xy} = 0$, $\sigma_x = -\sigma$ and hence it is the dangerous point at uni-axial compression of a wall or a tunnel.

3.2.6 General Relations for a Semi-Plane Under Vertical Load

Since in this case $\tau_{xy} = 0$ at $y = 0$ we have from (3.70)

$$zF''(z) = -\chi'(z), \chi(z) = F(z) - zF'(z). \tag{3.81}$$

So, instead of the second expression (3.68) and (3.69), (3.70) we receive:

$$\begin{aligned} \sigma_y - \sigma_x + 2i\tau_{xy} &= -4iyF''(z), \\ \sigma_y + i\tau_{xy} &= F'(z) + \overline{F'(z)} - 2iyF''(z), \end{aligned} \tag{3.82}$$

$$2G(u_x + iu_y) = \kappa F(z) - \overline{F(z)} - 2iy\overline{F'(z)}. \tag{3.83}$$

From (3.83) and the second (3.82) we have at $y = 0$

$$u_y = (\kappa + 1)\mathrm{Im}F(x_o)/G, \sigma_y = 2\mathrm{Re}F'(x_o)$$

and it is not difficult to notice that they differ only by a constant multiplier from border conditions

$$u_z = \mathrm{Re}w(x_o)/2G, \tau_y = -\mathrm{Im}w'(x_o)$$

(see relations (3.5)) for an anti-plane deformation. That allows to receive solutions of plane problems by replacing in similar results of longitudinal shear functions $w'(z)$ by $-2iF'(z)$.

3.2.7 Crack in Tension

Replacing in the first expression (3.14) τ_o by $i\sigma/2$ we get the solution for a plane with crack of length 2l perpendicular to tensile stresses σ in infinity (Fig. 3.16) in form

$$F'(z) = \sigma z/2\sqrt{z^2 - l^2},\ F(z) = \sigma\sqrt{z^2 - l^2}/2. \tag{3.84}$$

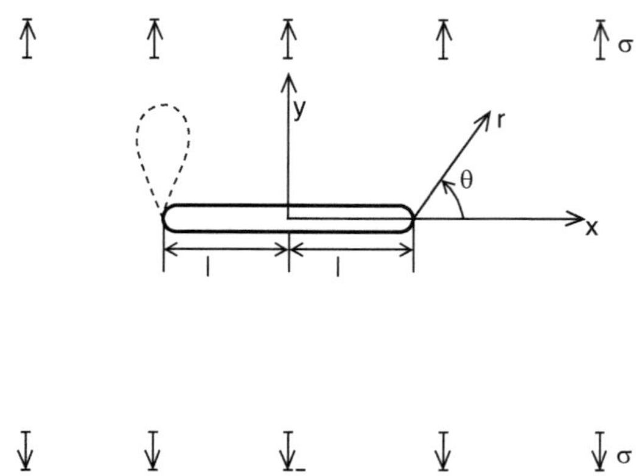

Fig. 3.16. Crack in tension

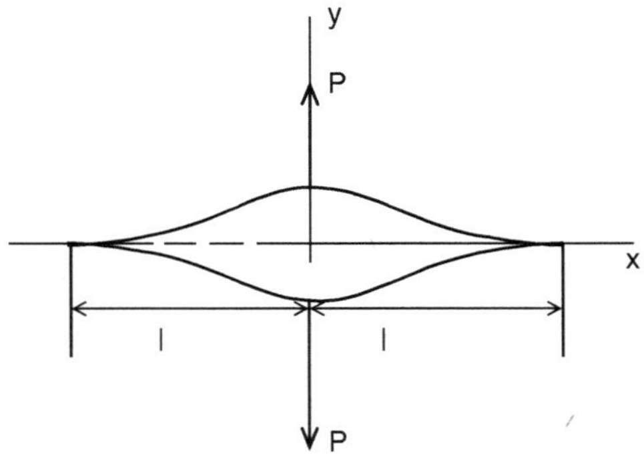

Fig. 3.17. Crack under concentrated forces

If $z \to \infty$ then $F'(z) = \sigma/2$, $F''(z) = 0$ and from (3.82), (3.84) we have $\sigma_x = \sigma_y = \sigma$ instead of $\sigma_x = 0, \sigma_y = \sigma$. So, the solution is wrong. But for the purpose of the further theory it is not important because stress σ_x does not influence the moment of a crack propagation beginning. Now since

$$F'' = -0.5\sigma l^2(z^2 - l^2)^{-3/2} \tag{3.85}$$

we find according to (3.82), (3.83) at $y = 0$, $x \leq$ /l/

$$\sigma_y = \tau_{xy} = \sigma_x = 0, \; u_y = (\sigma/4G)(k+1)\sqrt{l^2 - x^2}. \tag{3.86}$$

In a similar way the case in Fig. 3.17 can be considered where we have /17/

$$F'(z) = Pl/2\pi z\sqrt{z^2 - l^2}, \; F(z) = (P/2\pi)\cos^{-1} l/z$$

and it is an example of a stable crack because its length increases with a growth of the forces P.

3.2.8 Critical Strength

Now we find the energy due to the creation of a crack. According to the Clapeyron's theorem (3.17) and relation (3.86) we compute:

$$W = \sigma \int_{-l}^{l} /u_y/dx = (1+\kappa)\pi l^2\sigma^2/8G$$

and increment of the energy due to a propagation of the crack

$$dW = (1+\kappa)\pi l(\sigma_*)^2 dl/4G \tag{3.87}$$

must be equal to stretching resistance of the material $4\gamma_s dl$. That gives the critical value of σ

$$\sigma_* = 4\sqrt{G\gamma_s/\pi l(1+\kappa)} \tag{3.88}$$

which differs from (1.23) by a constant multiplier depending on the type of the problem (plane deformation or the same stress state) and the Poisson's ratio.

As in Sect. 3.1.4 we can find σ_* according to the relations of the asymptotic approach $z - l = e^{i\theta}$ which gives with a help of transformation equations like (2.55), (2.72)

$$u_r + iu_\theta = e^{-i\theta}(u_x + iu_y), \sigma_\theta - \sigma_r + 2i\tau_{r\theta} = (\sigma_y - \sigma_x + 2i\tau_{xy})e^{2i\theta} \tag{3.89}$$

as well as (3.82)–(3.84) and (2.65)

$$\begin{aligned}
&\sigma_\theta = (K_1/\sqrt{2\pi r})\cos^3(\theta/2),\ \sigma_r = (K_1/\sqrt{2\pi r})(1+\sin^2(\theta/2))\cos(\theta/2),\\
&\tau_{r\theta} = (K_1/2\sqrt{2\pi r})\sin\theta\cos(\theta/2),\ \tau_e = (K_1/2\sqrt{2\pi r})\sin\theta, && (3.90)\\
&u_r = (K_1/G)\sqrt{r/2\pi}(0.5(\kappa+1) - \cos^2(\theta/2))\cos(\theta/2), && (3.91)\\
&u_\theta = (K_1/G)\sqrt{r/2\pi}(0.5(1-\kappa) - \sin^2(\theta/2))\sin(\theta/2)
\end{aligned}$$

where

$$K_1 = \sigma\sqrt{\pi l}$$

is stress intensity coefficient for the first crack task. Constructing the equilibrium equation in its usual form

$$\int_0^{dl} /u_\theta(dl - r,\ \pi)/\sigma_\theta(r,\ 0)dr = 2\gamma_s dl \tag{3.92}$$

and using (3.90), (3.91) we find:

$$K_{1*} = 4\sqrt{G\gamma_s/(\kappa+1)} \tag{3.93}$$

and the strength condition is:

$$K_1 \leq K_{1*}.$$

There are some other approaches to a strength computation for brittle media. G. Barenblatt (see /17/) introduced a crack in a form of a beck (Fig. 3.18). His model gives strength value in 1.27 times higher than the Griffith's relation. Similar to that idea was introduced by N. Leonov and V. Panaciuk /19/. In their approach the crack begins to propagate when opening δ reaches its critical value δ_{cr} (Fig. 3.18). In this moment the critical strength can be computed as:

$$\sigma_* = \sqrt{E\sigma_{br}\delta_{cr}/\pi l} \tag{3.94}$$

where σ_{br} is a limit of brittle strength.

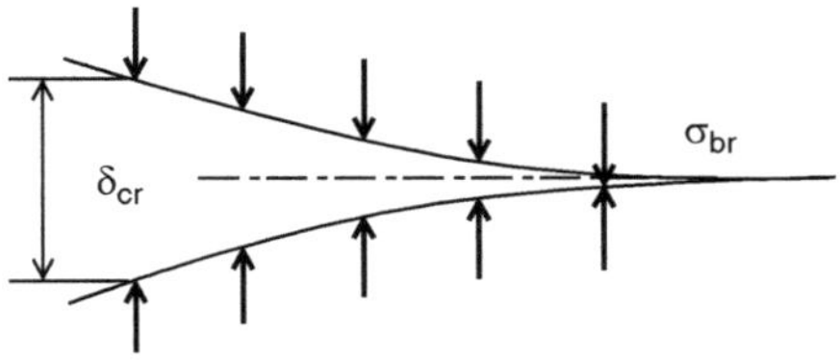

Fig. 3.18. Model of G. Barenblatt

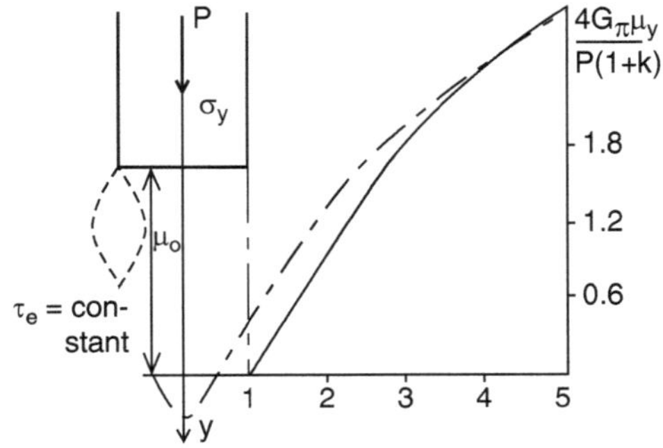

Fig. 3.19. Pressure of punch

3.2.9 Stresses and Displacements under Plane Punch

M. Sadowski solved this problem (Fig. 3.19) using the analogy method /20/. Replacing in the second relation (3.8) Q, ζ, $w'(\zeta)$ by P, z, $2iF'(z)$ respectively we receive:

$$F'(z) = -P/2\pi\sqrt{l^2 - z^2},\ F(z) = -(Pi/2\pi)\ln(z + \sqrt{z^2 - l^2}) + 2Gu_o. \quad (3.95)$$

We can easily notice that this result can be got from the first expression (3.84) after the consequent replacement of z, l, σ, by 1/z, 1/l, $-Pi/\pi l$ respectively.

With a help of (3.82), (3.83) we find a distribution of stresses (broken line for σ_y in Fig. 3.19) and displacement u_y (solid curves outside the punch) at $y = 0$ at $x < l$, $x > l$ respectively as:

$$u_y = u_o,\ \tau_{xy} = 0,\ \sigma_x = \sigma_y = -P/\pi\sqrt{l^2 - x^2}, \quad (3.96)$$

$$\sigma_x = \sigma_y = \tau_{xy} = 0,\ u_y = u_o - (P/2\pi G)(1 + \kappa)\ln(x/l - \sqrt{(x/l)^2 - 1}). \quad (3.97)$$

The computations show that diagram u_y outside the punch is near to that one for uniformly distributed load according to (3.51).

In a similar way as before we find with a help of (3.95), (3.82), (3.83) and (2.65) in the asymptotic approach

$$\begin{matrix}\sigma_r \\ \sigma_\theta\end{matrix} = -(P/\pi\sqrt{2rl})(1 \pm \cos^2(\theta/2))\sin(\theta/2),$$

$$\tau_{r\theta} = \left(P/2\pi\sqrt{2rl}\right)\sin\theta\sin(\theta/2),\ \tau_e = \left(P/2\pi\sqrt{2rl}\right)\sin\theta, \quad (3.98)$$

$$u_\theta = u_1 - (P/\pi G)\sqrt{r/2l}(0.5(\kappa+1) - \sin^2(\theta/2))\cos(\theta/2),$$
$$u_r = u_2 - (P/\pi G)\sqrt{r/2l}(0.5(\kappa-1) + \cos^2(\theta/2))\sin(\theta/2)$$

where u_1, u_2 – constants. It is easy to notice that τ_e in this task differs from that in the problem of crack (see the fourth relation (3.90)) by a constant multiplier. The condition τ_e = constant is shown by pointed line in the left part of Fig. 3.19 under the edge of the punch and the plastic zone must have this form.

From (3.98) we find critical load

$$P_* = 2\sqrt{2\gamma_s G/\pi l(k+1)}. \tag{3.99}$$

3.2.10 General Relations for Transversal Shear

In this case we have on axis x condition $\sigma_y = 0$ and from (3.68) we find after some simple transformations

$$\chi'(z) = -(2F'(z) + zF''(z)), \chi(z) = -(F(z) + zF'(z)). \tag{3.100}$$

Putting these expressions into (3.68), (3.69) we receive:

$$\sigma_y - \sigma_x + 2i\tau_{xy} = -4(F'(z) + iyF''(z)), \tag{3.101}$$
$$2G(u_x + iu_y) = \kappa F(z) + \overline{F(z)} - 2iy\overline{F'(z)}. \tag{3.102}$$

From (3.101) we have at $y = 0$

$$\tau_{xy} = -2\mathrm{Im}F'(x_o)$$

that is twice τ_y-value in the problem of the longitudinal shear in (3.5) and we can replace in the results of sub-chapter 3.1 $w'(z)$ by $2F'(z)$.

3.2.11 Rupture due to Crack in Transversal Shear

In this case (Fig. 3.20) $\tau_{xy}(\infty) = \tau$ and we derive from (3.14)

$$F'(z) = -i\tau z/2\sqrt{z^2 - l^2},\ F(z) = -0.5i\tau\sqrt{z^2 - l^2} \tag{3.103}$$

and according to (3.102) we find on axis x at $x < /l/$ and $x > /l/$ respectively

$$\tau_{xy} = u_y = \sigma_y = 0,\ \sigma_x = -2\tau x/\sqrt{l^2 - x^2},\ u_x = -\tau((\kappa+1)/2G)\sqrt{l^2 - x^2},$$
$$\sigma_x = \sigma_y = u_x = 0,\ \tau_{xy} = \tau x/\sqrt{x^2 - l^2},\ u_y = ((\kappa-1)/2G)\sqrt{x^2 - l^2} \tag{3.104}$$

and by the Clapeyron's theorem (3.17) as well as the energy balance

$$(1+\kappa)\pi l\tau_*^2 dl/4G = 4\gamma_s dl$$

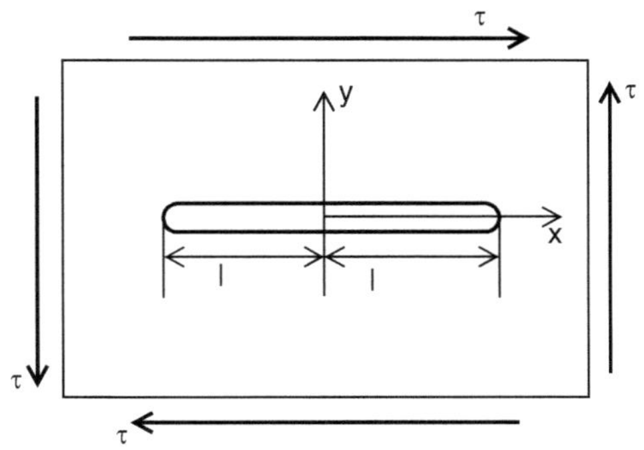

Fig. 3.20. Crack in transversal shear

we compute

$$\tau_* = 4\sqrt{G\gamma_s/\pi(\kappa+1)l}. \tag{3.105}$$

The same results can be received according to the asymptotic approach and (3.102), (3.103) as

$$\begin{aligned}
\sigma_r &= -(K_2/\sqrt{2\pi r})(2 - 3\cos^2(\theta/2))\sin(\theta/2),\\
\sigma_\theta &= -3(K_2/\sqrt{2\pi r})\cos^2(\theta/2)\sin(\theta/2),\\
\tau_{r\theta} &= (K_2/\sqrt{2r\pi})(1 - 3\sin^2(\theta/2))\cos(\theta/2),\\
\tau_e &= (K_2/2\sqrt{2\pi r})\sqrt{1 + 3\cos^2\theta},
\end{aligned} \tag{3.106}$$

$$\begin{matrix} u_r \\ u_\theta \end{matrix} = (K_2/2G)\sqrt{r/2\pi}x \begin{matrix} (-\kappa - 5 + 6\cos^2(\theta/2))\cos(\theta/2). \\ (-\kappa + 5 - 6\sin^2(\theta/2))\sin(\theta/2) \end{matrix} \tag{3.107}$$

Here $K_2 = \tau\sqrt{\pi l}$ – the stress intensity coefficient of the second crack task. Further computation follows that one for a crack in tension and we find a similar value (see also (3.93))

$$K_{2*} = 4\sqrt{G\gamma_s/(\kappa+1)} \tag{3.108}$$

and the strength condition

$$K_2 \leq K_{2*}.$$

Diagrams $\sigma_\theta/\sigma_{yi}$, σ_r/σ_{yi}, τ/σ_{yi} at $\tau_e = \sigma_{yi}/2$ as functions of θ are given in Fig. 3.21 by solid, broken and interrupted by points lines 1 respectively.

3.2.12 Constant Displacement at Transversal Shear

Using the analogy mentioned above we have from (3.8) at $\zeta = z$

$$F'(z) = -Q/2\pi\sqrt{z^2 - l^2},\ F(z) = 2Gu_o - (Q/2\pi)\ln\left(z/l + \sqrt{(z/l)^2 - 1}\right) \tag{3.109}$$

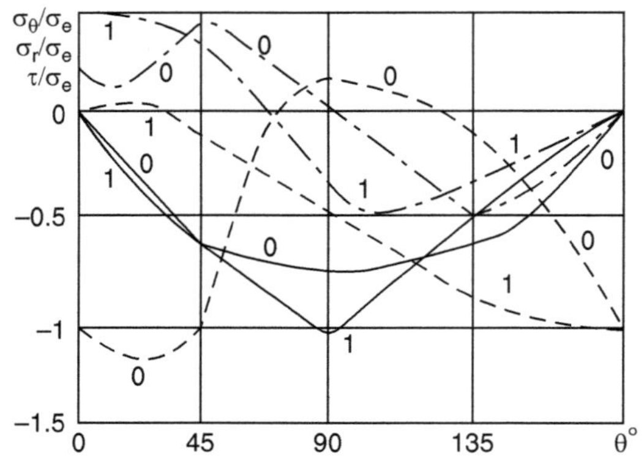

Fig. 3.21. Diagrams of stress distribution at crack ends in transversal shear

wherein Q is a resultant of τ_{xy} at $y = 0$, $-l < x < l$. From (3.101), (3.102), (3.109) we find on axis x at $x < /l/$ and $x > /l/$ respectively

$$u_x = u_o,\ u_y = \sigma_x = \sigma_y = 0,\ \tau_{xy} = Q/\pi\sqrt{l^2 - x^2},$$
$$\tau_{xy} = u_y = \sigma_y = 0,\ \sigma_x = -2Q/\pi\sqrt{x^2 - l^2},$$
$$u_x = u_o - (Q/2\pi G)(\kappa + 1)\ln(x/l + \sqrt{(x/l)^2 - 1}).$$

In the asymptotic approach we receive similarly to (3.106), (3.107) as

$$\begin{aligned}
&\sigma_r = -(Q/\pi\sqrt{2rl})(3\cos^2(\theta/2) - 1)\cos(\theta/2),\\
&\sigma_\theta = -3(Q/\pi\sqrt{2rl})\sin^2(\theta/2)\cos(\theta/2),\\
&\tau_{r\theta} = (Q/\pi\sqrt{2rl})(3\sin^2(\theta/2) - 2)\sin(\theta/2),\ \tau_e = (Q/2\pi\sqrt{2rl})\sqrt{1 + 3\cos^2\theta},\\
&u_r = u_1 - (Q/\pi G)\sqrt{r/2l}((\kappa + 1)/2 - 3\sin^2(\theta/2))\cos(\theta/2),\\
&u_\theta = u_2 - (Q/\pi G)\sqrt{r/2l}((\kappa - 1)/2 - 3\cos^2(\theta/2))\sin(\theta/2).
\end{aligned} \tag{3.110}$$

And we again can see that τ_e in (3.106), (3.110) differ by a constant multiplier. For critical stress we have expression (3.99) with Q instead of P.

3.2.13 Inclined Crack in Tension

By a combination of the solutions in Sects. 3.2.7, 3.2.11 a strength of a body with inclined crack in tension (Figure 3.22) can be studied. Supposing according to (2.72) $\sigma = p\sin^2\beta$, $\tau = 0.5p\sin 2\beta$ and seeking in the end of the crack main plane with $\theta = \theta_*$ L. Kachanov found in /17/ relation

$$\sin\theta_* + (3\cos\theta_* - 1)\cot\beta = 0 \tag{3.111}$$

according to which the crack must propagate in this direction. Some experiments confirm it.

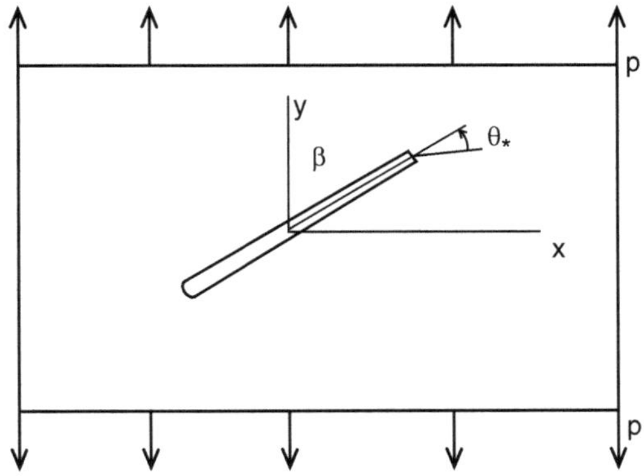

Fig. 3.22. Inclined crack in tension

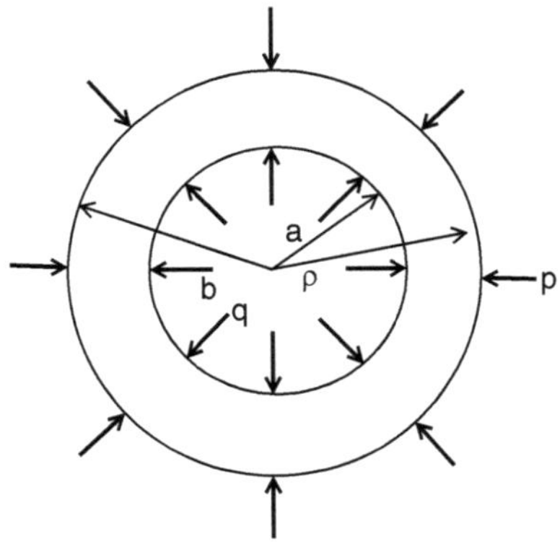

Fig. 3.23. Sphere under internal and external pressure

3.3 Axisymmetric Problem and its Generalization

3.3.1 Sphere, Cylinder and Cone Under External and Internal Pressure

For a sphere with internal a, current ρ and external b radii (Fig. 3.23) we use (2.80), (2.81) and the Hooke's law (2.17) at $\sigma_\theta = \sigma_\chi$ in the form

$$\sigma_\theta = E(\varepsilon_\theta + \nu\varepsilon_\rho)/(1-\nu-2\nu^2), \sigma_\rho = E(\varepsilon_\rho(1-\nu)+2\nu\varepsilon_\theta)/(1-\nu-2\nu^2). \quad (3.112)$$

Putting (2.81) into (3.112) and the result – in (2.80) we get on a differential equation for $u_\rho \equiv u$

$$d^2u/d\rho^2 + 2du/\rho d\rho - 2u/\rho^2 = 0$$

with an obvious integral

$$u = A/\rho^2 + B\rho. \quad (3.113)$$

Now we determine strains from (2.80), stresses – by (3.112) and constants – according to boundary conditions $\sigma_\rho(a) = -q$, $\sigma_\rho(b) = -p$. As a result we have

$$\begin{aligned}\sigma_\theta &= (qa^3(2\rho^3 + b^3) - pb^3(2\rho^3 + a^3))/2\rho^3(b^3 - a^3),\\ \sigma_\rho &= (qa^3(\rho^3 - b^3) + pb^3(a^3 - \rho^3))/\rho^3(b^3 - a^3).\end{aligned} \tag{3.114}$$

The strains and the displacements can be found according to the Hooke's law and expression (2.80).

In a similar way the stress distribution in a tube can be analysed. To change the method we use here potential function Φ (see Sect. 2.4.3) since the problem is a plane one. The biharmonic equation (2.74) in this case becomes

$$d(rd(d(rd\Phi/dr)/rdr)/dr)/rdr = 0$$

with a very simple solution that with a help of (2.75) and boundary conditions like that for the sphere (with replacement in them ρ by r) gives

$$\begin{aligned}\sigma_r &= (a^2b^2(p - q)/r^2 + qa^2 - pb^2)/(b^2 - a^2),\\ \sigma_\theta &= (qa^2 - pb^2 + a^2b^2(q - p)/r^2)/(b^2 - a^2).\end{aligned} \tag{3.115}$$

Let us now consider a cone (Fig. 3.24) for which we use spherical coordinates (Fig. 2.10) and supposition $\tau_{\rho\theta} = \tau_{\rho\chi} = \tau_{\chi\theta} = \varepsilon_\rho = \gamma_{\rho\theta} = \gamma_{\rho\chi} = \gamma_{\chi\theta} = 0$ (as in a cylinder). Other components do not depend on ρ, θ. Above that $u_\rho = u_\theta = 0$ and $u_\chi = \rho u(\chi)$. In coordinates θ, χ the first equation (2.77) takes the form

$$d\sigma_\chi/d\chi + (\sigma_\chi - \sigma_\theta)\cot\chi = 0 \tag{3.116}$$

and expressions (2.79) give

$$u = C/\sin\chi, \varepsilon_\theta = -\varepsilon_\chi = C\cos\chi/\sin^2\chi. \tag{3.117}$$

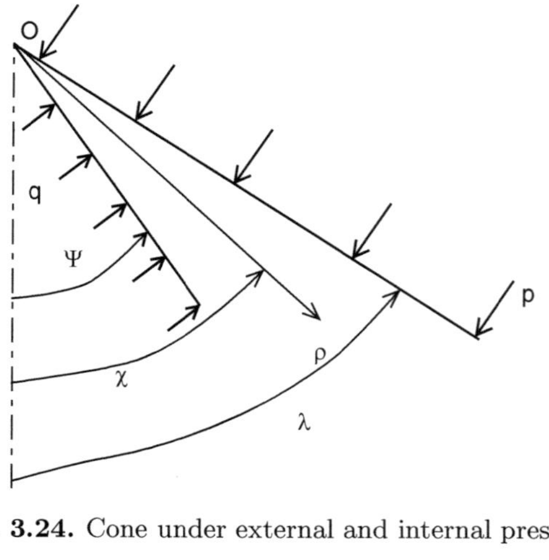

Fig. 3.24. Cone under external and internal pressure

Now we use the Hooke's law (2.17) that leads to relation

$$\sigma_\theta - \sigma_\chi = 4GC\cos\chi/\sin^2\chi$$

and from (3.116) – to

$$\sigma_\chi = D - 2GC(\cos\chi/\sin^2\chi + \ln\tan(\chi/2)). \qquad (3.118)$$

Constants C, D have to be determined from border demands $\sigma_\chi(\psi) = -q$, $\sigma_\chi(\lambda) = -p$. As a result we derive finally

$$\begin{matrix}\sigma_\theta \\ \sigma_r\end{matrix} = -q + (q-p)(\cos\psi/\sin^2\psi \pm \cos\chi/\sin^2\chi - \ln(\tan(\chi/2)/\tan(\psi/2)))/A$$

where

$$A = \cos\psi/\sin^2\psi - \cos\lambda/\sin^2\lambda + \ln(\tan(\psi/2)/\tan(\lambda/2)). \qquad (3.119)$$

From expression (3.117) we find deformations and displacement for incompressible body as follows

$$\varepsilon_\theta = (p-q)\cos\chi/2GA\sin^2\chi = -\varepsilon_\chi, u = (p-q)/2AG\sin\chi.$$

This solution can model a behaviour of a volcano. When ψ, λ, χ tend to zero we get the Lame's relations for the tube that were derived above.

The theory of this section can be used for an appreciation of the strength of different voids in a medium.

3.3.2 Boussinesq's Solution and its Generalization

Stresses in Semi-Space Under Concentrated Load

If an external concentrated force F acts vertically in point O (Fig. 2.10) on a semi-infinite solid the stresses in point N are /5/

$$\begin{aligned}\sigma_z = -3Fz^3/2\pi\rho^5, \sigma_r = F((1-2\nu)(\rho - z)/\rho r^2 - 3r^2z/\rho^5)/2\pi,\\ \sigma_\theta = F(1-2\nu)(zr^2 + z\rho^2 - \rho^3)/2\pi r^2\rho^3, \tau_{rz} = -3Frz^2/2\pi\rho^5.\end{aligned} \qquad (3.120)$$

These relations are known as the Bousinesq's solution for axisymmetric problem published in 1889 and they are similar to the Flamant's expressions in Sect. 3.2.3 for plane one. Using (2.72) we compute

$$\begin{aligned}\sigma_\rho &= F((1-2\nu)(1-z/\rho) - 3z/\rho)/2\pi\rho^2,\\ \sigma_\chi &= Fz^2(1-2\nu)(1-z/\rho)/2\pi r^2\rho^2,\\ \tau_{\rho\chi} &= Fz(1-2\nu)(1-z/\rho)/2\pi r\rho^2\end{aligned} \qquad (3.121)$$

and we can see that only for incompressible material ($\nu = 0.5$) directions ρ, χ are main ones and $\sigma_\chi = \sigma_\theta = 0$.

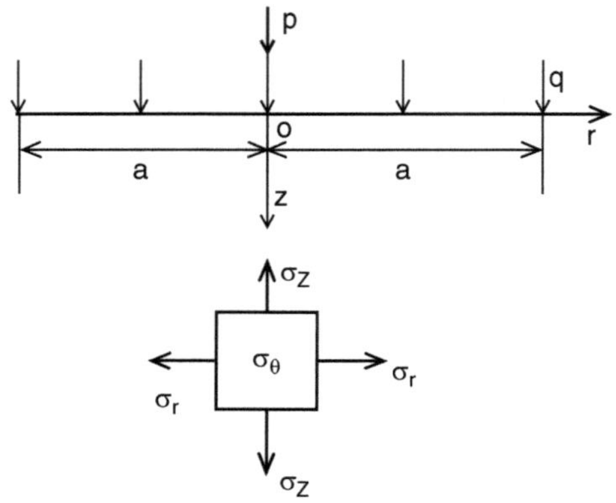

Fig. 3.25. Stresses under uniformly distributed load in circle

Stresses Under Distributed Load

Using the superposition method we can find stresses under any load. As the first example we consider a circle of radius a under uniformly distributed forces q. Firstly we study stresses along axis z where we have /5/

$$\sigma_z = q(z^3(a^2 + z^2)^{-3/2} - 1). \tag{3.122}$$

In the same manner stresses σ_r, σ_θ (Fig. 3.25) can be found as

$$\sigma_\theta = \sigma_r = q\left(-1 - 2\nu + 2(1+\nu)z/\sqrt{a^2 + z^2} - z^3(a^2 + z^2)^{-3/2}\right)/2. \tag{3.123}$$

Particularly in point O we have

$$\sigma_z = -q, \sigma_r = \sigma_\theta = -q(1 + 2\nu)/2.$$

The maximum shearing stress can be easily computed according to (2.10), (3.122), (3.123) as follows

$$\tau_e = q(0.5(1 - 2\nu) + (1 + \nu)z/\sqrt{a^2 + z^2} - 3z^3(a^2 + z^2)^{-3/2})/2. \tag{3.124}$$

This expression has its maximum at $z_* = a\sqrt{(1+\nu)(7-2\nu)}$ and it is

$$\max\tau_e = q(0.5(1 - 2\nu) + 2(1 + \nu)\sqrt{2(1 + \nu)/9})/2. \tag{3.125}$$

For example if $\nu = 0.3$ then $z_* = 0.64a$ and $\max\tau_e = 0.33q$.

An interesting case takes place for a circular punch and Boussinesq gave the solution in a form similar to (3.95) as

$$q = P/2\pi a\sqrt{a^2 - r^2} \tag{3.126}$$

where P is a resultant of loads q. The least value of q is in the centre: $q_{min} = P/2\pi a^2$. Diagram q(r) is given in Fig. 3.26 by broken line and as we can see the stresses are very high at r = a (similar to other problems of punches and cracks in plane problem). In reality plastic strains appear at the edges, redistribution of stresses occurs and q(r) diagram has a form of the solid curve in the figure.

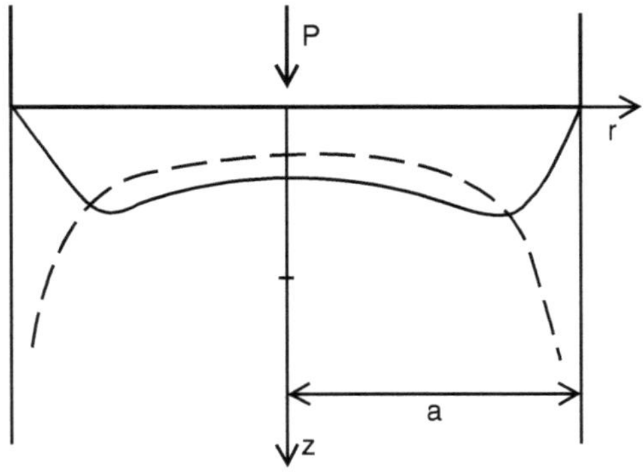

Fig. 3.26. Distribution of stresses under circular punch

Stresses Under Rectangles

The linear dependence of stresses on displacements allows to use the superposition principle for finding stresses at different loadings. To realize that we rewrite the first relation (3.120) for the stress in a point with coordinates z, r (Fig. 2.10) as

$$\sigma_z = K_\sigma F/z^2. \tag{3.127}$$

Here (in this section compressive stresses are taken positive)

$$K_\sigma = 3/2\pi(1 + r^2/z^2)^{5/2}$$

is a coefficient the values of which are given in special tables (see Appendix B).

When several (n) forces act then stress σ_z is computed as follows

$$\sigma_z = \left(\sum_{i=1}^{n} K_{\sigma i} F_i\right)/z^2$$

where factors $K_{\sigma i}$ are taken as the functions of ratio r_i/z and r_i is the distance from the studied point to the direction of a F_i action. This method can be applied to a case of distributed load when we lay out a considered area on separate parts and compute the resultant for each of them.

The special particularly important case takes place when we have uniformly distributed load over a rectangle. Here we lay out the whole area on separate rectangles and find the stress in the common for them point as a sum of the stresses in each of the parts. The following options can be met (Fig. 3.27):

1. Point M is on a border of the rectangle (Fig. 3.27, a) and we summarize stresses due to loads in rectangles abeM and Mecd,
2. Point M is inside a rectangle (Fig. 3.27, b) and we summarize the stresses from the action of the load in rectangles Mhbe, Mgah, Mecf and Mfdg,

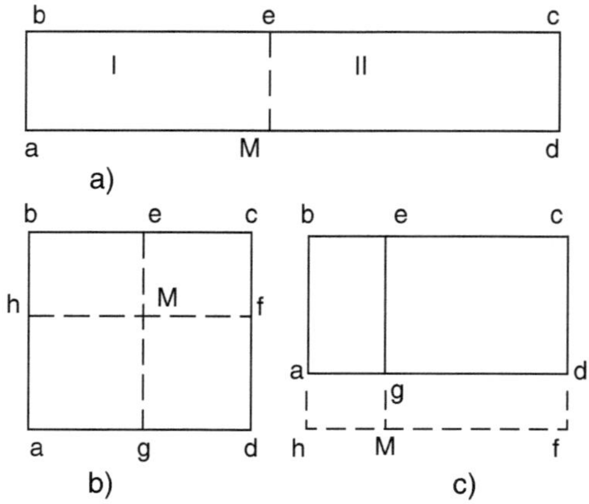

Fig. 3.27. Uniformly distributed load over rectangle

3. Point M is outside a rectangle (Fig. 3.27, c) and we summarize the stresses from the action of a load in rectangles Mhbe and Mecf and subtract that in rectangles Mhag and Mgdf.

The determination of stresses is fulfilled with the help of special tables according to relation

$$\sigma_z = K'q \tag{3.128}$$

where factor K' is given in the function of ratios $m = l/b$ – relative length and $n = z/b$ – relative depth (see Appendix C). q is an intensity of the loads. For example in the case 1 we have

$$\sigma_z = q((K')_I + (K')_{II}). \tag{3.129}$$

Displacements in Massif

We begin with the case of concentrated force F when we have according to the Hooke's law on the surface $z = 0$ /5/

$$u_r = -(1-2\nu)(1+\nu)F/2\pi Er, u_z \equiv S = F(1-\nu^2)/\pi Er. \tag{3.130}$$

In other cases we use the superposition method. E.g. for a circle of radius a under uniformly distributed load q we write for a point outside it

$$u_z = 4q(1-\nu^2)r(L(a/r) - (1-a^2/r^2)K(a/r))/\pi E \tag{3.131}$$

where K(a/r), L(a/r) are full elliptic integrals of the first and the second kind. They can be calculated with a help of special tables. For the settling of the external circumference ($r = a$) we receive

$$u_z = 4(1-\nu^2)qa/\pi E \tag{3.132}$$

and in points inside the circle the displacement is

$$u_z = 4(1 - \nu^2)qaL(a/r)/\pi E. \tag{3.133}$$

The highest displacement is in the centre of the circle as

$$\max u_z = 2(1 - \nu^2)qa/E$$

and it is easy to prove that $\max u_z/u_z(a) = \pi/2$. Now we find a mean displacement as

$$\mathrm{mean} u_z = (P/\pi a^2)\int_0^a 2\pi u_z r dr = 0.54F(1 - \nu^2)/\pi E$$

and it is near to the displacement under a circular punch

$$u_z = 0.5F(1 - \nu^2)/\pi E. \tag{3.134}$$

A similar situation takes place for a square with sides 2a loaded by uniformly distributed forces q. In this case

$$\max u_z = 8qa\ln(\sqrt{2} + 1)(1 - \nu^2)/\pi E = 2.24qa(1 - \nu^2)/E. \tag{3.135}$$

In corners $u_z = 0.5 \max u_z$ and an average u_z is equal to $1.9qa(1 - \nu^2)/E$. The same computations were made for rectangles with different ratios of h/b. The results are represented in a form

$$u_z = m_o q(1 - \nu^2)/E. \tag{3.136}$$

The values of m_o are given in table here as a function of the sides ratio h/b.

h/b	circle	1	1.5	2	3	5	10	100
m_o	0.96	0.95	0.94	0.92	0.88	0.82	0.71	0.37

Approximate Methods of Settling Computations

In practice some approximate approaches are used for a computation of a settling. One of them is a method of a summation "layer by layer". Here the hypothesis is taken that a lateral expansion is absent or in other words that the dependence of stresses on porosity is compressive (see Sect. 1.4.2). It is also supposed that a decrease of σ_z with a depth subdues to Boussinesq's solution (Fig. 3.28). The whole settling is calculated as a sum of displacements of elementary layers /10/

$$S = \beta_o \sum_{i=1}^{n} \sigma_{zi} h_i / E_i. \tag{3.137}$$

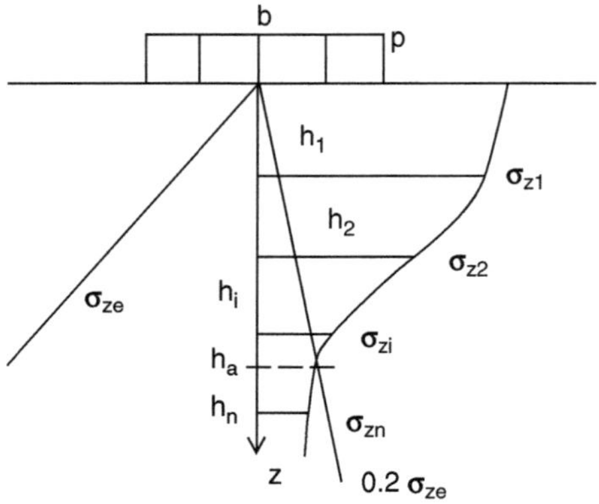

Fig. 3.28. Approximate computation of settling

Here β_o is a dimensionless coefficient equal usually to 0.8, h_i, E_i – a thickness and a modulus of deformation of i-layer, σ_{zi} is computed according to the first relation (3.120) for the middle of the layer, h_n is taken for a layer where the settling is small. In Russia $\sigma_{zn} = 0.2\sigma_{ze}$ where σ_{ze} are stresses from earth's self-weight (see Sect. 2.4.1). If this layer has $E < 5\,\text{MPa}$ it is included in sum (3.128). For hydro-technical structures with big width b (Fig. 3.28) condition $\sigma_z > 0.5\sigma_{ze}$ is usually taken.

Another approach to the solution of this problem gave N. Cytovich /3/ who proposed to take into account some lateral expansion of the soil and an influence of a footing size (see Fig. 1.6). He introduced the so-called equivalent layer h_s which exposes the same settling as in the presence of a lateral expansion:

$$h_s = (1 - \nu^2)\eta b. \tag{3.138}$$

Here parameter η considers a form and a rigidity of a footing with width b.

When a foundation has a form of a rectangle the method of corner points is applied similar to that for the calculation of stresses.

3.3.3 Short Information on Bending of Thin Plates

General Equations for Circular Plates

A plate is considered to be thin when the ratio of its thickness to the minimum dimension in plane L satisfies the condition $0.2 > h/L > 0.0125$. This problem is studied in special courses and comparatively simple theory exists for axisymmetric plates. Differential equation of their element (Fig. 3.29) is

$$M_\theta - d(M_r r)/dr = Qr. \tag{3.139}$$

Here M_r, M_θ are radial and tangential bending moments, Q – transversal (shearing) force which can be computed according to an equilibrium condition

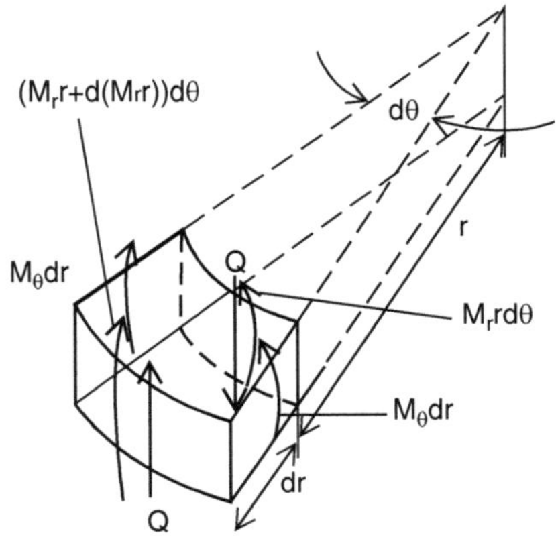

Fig. 3.29. Element of circular plate

of a middle part of the plate with radius r. In the case of uniformly distributed load q it is

$$Q = 0.5qr. \tag{3.140}$$

For a plate of radius R at q = constant in a circle of diameter $2r_o$ integration of (3.139) with consideration of (3.140) gives equation

$$RM_r(R) + q(r_o)^3/6 + q(r_o)^2(R - r_o)/2 = \int_0^R M_\theta dr. \tag{3.141}$$

Replacing in (3.141) $q(r_o)^2$ by F/π and supposing $r_o = 0$ we come to the case of concentrated force F in the centre of the plate as

$$FR/2\pi = \int_0^R M_\theta dr - RM_r(R). \tag{3.142}$$

At $R = r_o$ we have from (3.141) the solution for a plate under uniformly distributed pressure q in form

$$RM_r(R) + qR^3/6 = \int_0^R M_\theta dr. \tag{3.143}$$

Similarly some other cases can be considered.

Ultimate State of Circular Plates

Now we find according to the first Gvozdev's theorem the ultimate state of the plate. If its edges are freely supported we must put in (3.142), (3.143)

$M_r(R) = 0$, $M_\theta = M_*$ (where $M_* = \sigma_{yi}h^2/4$ – see expression (1.24)) which gives

$$F_* = 2\pi M_*, q_* = 6M_*/R^2. \tag{3.144}$$

Comparing these F_* and q_* to (1.28) and the first (1.31) we see that they coincide and hence are rigorous. If the edges of the plate are fixed we must put in (3.142), (3.143) – $M_r(R) = M_\theta = M_*$. That leads to ultimate values

$$F_* = 4\pi M_*, q_* = 12M_*/R^2 \tag{3.145}$$

which coincide with relation (1.30) and the second expression (1.31) respectively. Therefore they are also exact. In the similar way some other different cases of an axi-symmetric load can be considered.

Ultimate State of Square Plates

The bending of rectangular plates are usually studied in double trigonometric series. For example for a square 2R × 2R in plane loaded by uniformly distributed pressure q with origin of coordinate system x, y in one of its corners we have

$$M_x/M_y = (16qR^2/\pi^4) \sum_{m=1}^{\infty} \sum_{n=1}^{\infty} {}^{m^2+\nu n^2}_{n^2+\nu m^2} \left(x \left(\frac{\sin m\pi x}{2R} \right) \left(\frac{\sin m\pi y}{2R} \right) \right) /mn(m^2+n^2)^2 (m, n = 1, 3, ..). \tag{3.146}$$

Taking only the first member of the series we find for maximum moments (in the centre of the plate)

$$\max M_x = \max M_y = (1+\nu)qR^2/6$$

which give the ultimate load as

$$q_* = 6M_*/(1+\nu)R^2. \tag{3.147}$$

Another simple solution can be received when we use differential equation of an element of the plate in form

$$\partial^2 M_x/\partial x^2 + 2\partial^2 M_{xy}/\partial x \partial y + \partial^2 M_y/\partial y^2 = -q \tag{3.148}$$

where M_{xy} is the moment of a torsion. Taking approximately for the moments expressions that satisfy the border demands (here the origin of the coordinate system is in the centre of the plate) $M_x = C(R^2 - x^2)$, $M_y = C(R^2 - y^2)$, $M_{xy} = 0$ and putting them into (3.148) we find $C = q/4$ and hence

$$q_* = 4M_*/R^2$$

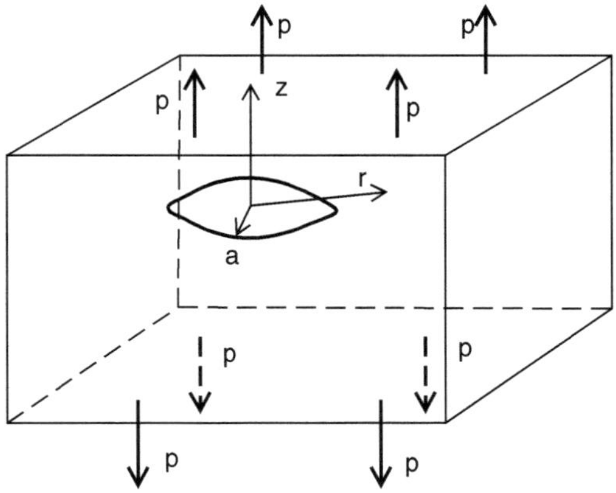

Fig. 3.30. Circular crack in tension

which coincides with (3.147) for incompressible material.

Taking into account (3.147) and the first relation (1.31) we find the following limits for the ultimate load

$$6M_*/R^2(1+\nu) \leq q_* \leq 6M_*/R^2. \quad (3.149)$$

We can see that the q_*-value is rigorous for $\nu = 0$.

3.3.4 Circular Crack in Tension

Here (Fig. 3.30), (2.76) are valid and we seek solution in the following form /17/

$$u_z = -\partial^2\Phi/2G\partial r\partial z, u_z = (2(1-\nu)\Delta\Phi - \partial^2\Phi/\partial z^2)/2G$$

where Φ is a function of z and r. Putting these expressions into (2.76) and then strains – into the Hooke's law we get the stresses as

$$\sigma_r = \partial(\nu\Delta\Phi - \partial^2\Phi/\partial r^2)/\partial z, \sigma_\theta = \partial(\nu\Delta\Phi - \partial\Phi/r\partial r)\partial z,$$

$$\sigma_z = \partial((2-\nu)\Delta\Phi - \partial^2\Phi/\partial z^2)\partial z, \tau_{rz} = \partial((1-\nu)\Delta\Phi - \partial^2\Phi/\partial z^2)/\partial r.$$

These values satisfy static equations and putting them into compatibility relation (2.76) we find biharmonic law for Φ.

Using the the Henkel's transformations we get expressions for the displacement and stress that at p = constant, $\rho = r/a$ and $z = 0$ are

$$\begin{aligned} u_z(\rho, 0) &= 4(1-\nu^2)pa\sqrt{1-\rho^2})/\pi E\,(\rho \leq 1),\\ \sigma_z &= 2p\left(1/\sqrt{\rho^2-1} - \sin^{-1}(1/\rho)\right)(\rho > 1). \end{aligned} \quad (3.150)$$

Since near the crack edges the first member in brackets is much higher than the second one the solution is somewhat similar to that (3.126) for the circular punch.

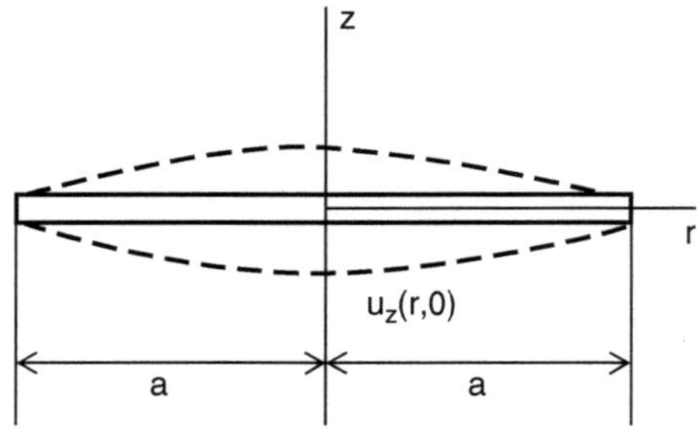

Fig. 3.31. Deformation of crack

From Fig. 3.31 where the curve $u_z(\rho)$ according to the first (3.150) is shown by broken lines we can see that deformed crack is an ellipsoid. Stress σ_z has the same peculiarity as in similar problems at the longitudinal shear and plane deformation. Using the expression for the work at crack propagation

$$W = -2p\pi a^2 \int_0^1 u_z(\rho, 0) d\rho,$$

relation (3.150) for $u_z(\rho, 0)$ and equality

$$dW = 2\pi\gamma_s a da$$

as in Sects. 3.1.4, 3.2.8 and 3.2.11 we find

$$K_{o*} = \sqrt{\gamma_s E/(1-\nu^2)} = 2p\sqrt{a}/\sqrt{\pi}.$$

The problem of non-uniformly distributed forces is solved in the same manner. The task of two cracks in distance $z = \pm z_1$ is also studied in a space and in a cylinder.

4

Elastic-Plastic and Ultimate State of Perfect Plastic Bodies

4.1 Anti-Plane Deformation

4.1.1 Ultimate State at Torsion

Although exact elastic solutions at torsion are known only for some cross-sections the ultimate state can be found for any problem because in this case we should consider only two equations for two unknowns (Fig. 2.8) – condition $\tau_e = \tau_{yi}$ together with static law (2.48). It can be satisfied if we take

$$\tau_x = \partial w/\partial y, \tau_y = -\partial w/\partial x \tag{4.1}$$

and from 2.52 we find

$$(\partial w/\partial x)^2 + (\partial w/\partial y)^2 = (\tau_{yi})^2$$

or

$$/\mathrm{grad} w/ = \tau_{yi} = \text{constant}. \tag{4.2}$$

Here the gradient w(x, y) is the maximum slope of that function which can be interpreted as a sand heap with angle of repose equal to $\tan^{-1}\tau_{yi}$. Expression (4.2) means that the distance between lines in which τ_{yi} acts are constant and it allows to compute an elementary moment of torsion as (Fig. 4.1)

$$dM_* = \tau_{yi} p dp ds = \tau_{yi} 2 dp dA$$

where A is an area under curve w = constant and p–perpendicular to it from a pole. Summarizing dM_* we find the ultimate moment as

$$M_* = 2V. \tag{4.3}$$

Here V is volume of the heap.

Relation (4.3) opens the way for finding the ultimate load experimentally. For some sections M_*-value can be calculated. In the case of a circle with radius R e.g. we have

$$M_* = 2\pi R^3 \tau_{yi}/3.$$

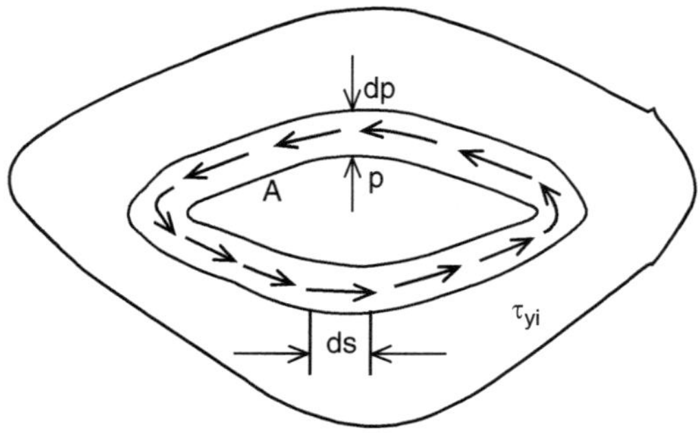

Fig. 4.1. Computation of torsion moment

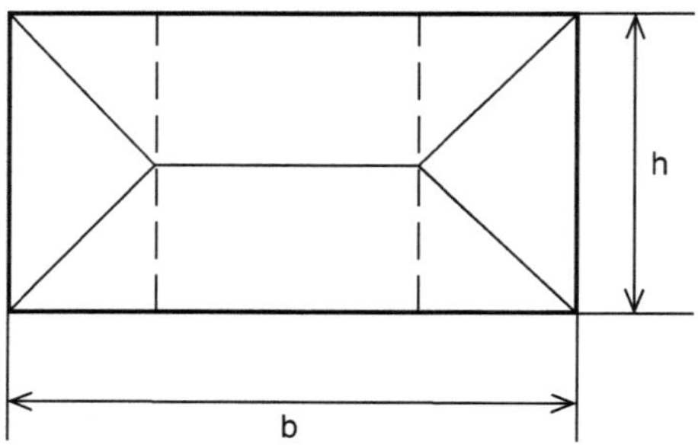

Fig. 4.2. Ultimate moment for rectangle

For a rectangle (Fig. 4.2) we compute in a similar way

$$M_* = h^2(3b - h)\tau_{yi}/6.$$

From this expression we have as particular cases M_*-values for a long strip and a square:

$$M_* = bh^2\tau_{yi}/2, M_* = h^3\tau_{yi}/3.$$

Relations above can be used for τ_{yi} determination by the torsion tests of plastic materials including some soils.

4.1.2 Plastic Zones Near Crack and Punch Ends

When plastic strains appear the value of τ_* (see Sect. 3.1.4) falls as G decreases in many times. There are several solutions for a perfect plastic body. We follow here the approaches of Rice /21/ for small plastic zones where relations (3.19) are valid. As we mentioned there the condition τ_e = constant gives a circumference and we can suppose that the plastic zone is a circle (Fig. 4.3) with a radius R_o that can be found from boundary conditions for stresses and displacement u_z on the border between elastic and plastic districts.

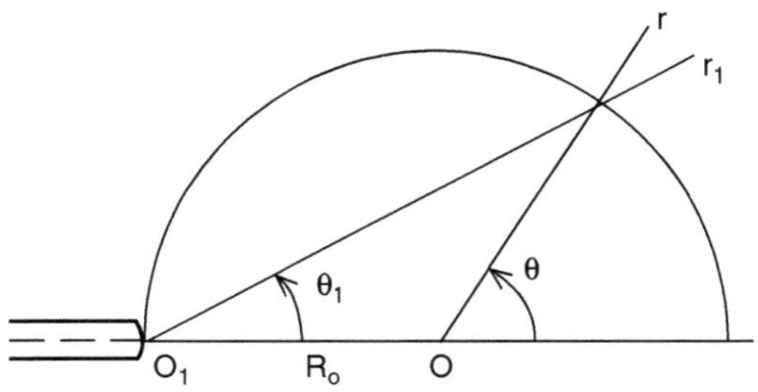

Fig. 4.3. Plastic zone near crack end

According to relations of (2.55) type we find with a help of (3.19)

$$\tau_{r1} = 0,\ \tau_{\theta 1} = \tau_o \sqrt{l/2r}.$$

Taking τ_e as τ_{yi} we receive R_o as follows

$$R_o = (\tau_o)^2 l/2(\tau_{yi})^2. \tag{4.4}$$

Now we consider the displacements and from elastic part of the body we have according to (4.4) and the first expression (3.19)

$$u_z = (\tau_o/G)\sqrt{2lR_o}\sin\theta/2. \tag{4.5}$$

Then from the second expression (2.57) we find on the circumference starting from the plastic zone at $r_1 = 2R_o\cos\theta_1$

$$u_z = (\tau_{yi}/G)2R_o \int_0^{\theta_1} \cos\theta_1 d\theta_1$$

or after computations

$$u_z = 2(\tau_{yi}/G)R_o\sin(\theta/2). \tag{4.6}$$

Taking into account equality $\sin\theta_1 = \theta/2$ and comparing (4.6) to (4.5) we get on expression (4.4) that is the same value of R_o. Lastly we determine the displacement in the end of the crack in form

$$\delta = 2u_z(R_o, \pi) = 2l(\tau_o)^2/G\tau_{yi}.$$

In the same manner the problem of the strip's longitudinal movement can be studied. Using expressions (2.58) and (3.10) we find on the circumference $r = R_o$:

$$\tau_{\theta 1} = 0,\ \tau_{r1} = \tau_e = Q/\pi\sqrt{2R_o l}.$$

Supposing $\tau_{r1} = \tau_{yi}$ we find the radius R_o of the plastic zone as

$$R_o = Q^2/2\pi^2 l(\tau_{yi})^2. \tag{4.7}$$

Now if we replace in the third relation (3.10) r by R_o this equation is valid on the border from the elastic side. And from the plastic region we have according to the first (2.57)

$$u_z = u_o - \tau_{yi} r_1/G$$

or

$$u_z = u_o - 2R_o(\tau_{yi}/G)\cos(\theta/2).$$

Comparing this expression to the first relation (3.10) at $r = R_o$ we find again (4.7). The position of the circumference's centre will be determined in the next chapter.

4.2 Plane Deformation

4.2.1 Elastic-Plastic Deformation and Failure of Slope

Stresses in Wedge

As was told in Sect. 3.2.1 the maximum shearing stress τ_e in the cases $\lambda > \pi/4$ reaches its maximum at $\theta = 0$ and there the first residual strains appear when load p is

$$p_{yi} = 2\tau_{yi}(2\lambda\cos 2\lambda - \sin 2\lambda)/(\cos 2\lambda - 1).$$

At $p > p_{yi}$ we have in Fig. 3.5 plastic zone BOC as well as two elastic districts AOB and COD for which the solution of Sect. 3.2.1 is valid in form

$$\begin{aligned} \tau &= C_1 + C_2\cos 2\theta + C_3\sin 2\theta, \\ \begin{matrix}\sigma_\theta \\ \sigma_r\end{matrix} &= C_4 - 2C_1\theta \pm C_3\cos 2\theta \pm (-C_2\sin 2\theta). \end{aligned} \tag{4.8}$$

In the plastic zone we take the same condition as on the straight line $\theta = 0$ at $p = p_{yi}$ that is $\tau = \tau_{yi}$ and from (3.21) we find with consideration of demand $\sigma_r(0) = \sigma_\theta(0) = -p/2$

$$\sigma_r = \sigma_\theta = -p/2 - 2\tau_{yi}\theta.$$

Constants $C_i(i = 1\text{–}4)$ and τ_{yi} should be found from the compatibility equations for stresses at $\theta = \pm\upsilon$ and boundary conditions $\tau(-\lambda) = \sigma_\theta(-\lambda) = 0$ and $\tau(\lambda) = 0, \sigma_\theta(\lambda) = -p$ on lines OA and OD respectively. As a result we have

$$\begin{aligned} \tau_{yi} &= C_o(\cos 2(\lambda - \upsilon) - 1), \\ \sigma_r = \sigma_\theta &= 2C_o\theta(1 - \cos 2(\lambda - \upsilon)) - p/2 \end{aligned} \tag{4.9}$$

where

$$C_o = 0.5p/(-2\upsilon + 2\lambda\cos 2(\lambda - \upsilon) - \sin 2(\lambda - \upsilon)) \qquad (4.10)$$

From (4.9), (4.10) the link between p/τ_{iy} and υ for different λ may be found. In elastic districts AOB, COD at upper and lower signs before υ are the stresses are:

$$\tau = C_o(\cos 2(\lambda - \upsilon) - \cos(\theta \pm \upsilon)),$$

$$\begin{matrix}\sigma_\theta \\ \sigma_r\end{matrix} = C_o(-(\pm(-2\upsilon)) - 2\theta\cos 2(\lambda - \upsilon) \pm \sin 2(\theta \pm \upsilon)) - p/2.$$

At $\lambda - \upsilon = \pi/4$ and $\tau = \tau_{yi}$ we have from (4.9), (4.10) the ultimate load as follows

$$p_u = 2\tau_{yi}(2\lambda - \pi/2 + 1) \qquad (4.11)$$

and it is interesting to notice that if we take the solution that is recommended in /18/ by V. Sokolovski for the case $\lambda \leq \pi/4$ which in our case gives the smaller load at $\pi/2 > \lambda > \pi/4$ as

$$(p_u)' = 2\tau_{yi}(\sin 2\lambda - (\pi/2 - 1)\cos 2\lambda).$$

However the last relation predicts a fall of the ultimate load with an increase of λ as a whole (e.g. $(p_u)'(\pi/2) = 1.14\tau_{yi}$) that contradicts a real behaviour of foundations.

Displacements in Wedge

In order to find displacements we use expressions (2.69), (2.66) in which $m = 1, \Omega = 1/G$ and indices x, y are replaced by r, θ respectively. As a result we have in districts AOB and COD at upper and lower signs before υ consequently

$$\begin{aligned}\begin{matrix}\varepsilon_\theta \\ \varepsilon_r\end{matrix} &= (-0.5p(1 - 2\nu) + C_o((1 - 2\nu)(-(\pm 2\upsilon) - 2\theta\cos 2(\lambda - \upsilon) \\ &\quad \pm \sin 2(\theta \pm \upsilon)))/2G, \\ \gamma &= C_o(\cos 2(\lambda - \upsilon) - \cos 2(\theta \pm \upsilon))/G, \\ u_r &= C_o r(-0.5p(1 - 2\nu)(-(\pm 2\upsilon) - 2\theta\cos 2(\lambda - \upsilon)) - \sin 2(\theta \pm \upsilon))/2G, \\ u_\theta &= C_o(4(1 - \nu)\ln(r/d)\cos 2(\lambda - \upsilon) - \cos 2(\theta - \upsilon))/2G.\end{aligned}$$

where d has the same meaning as in relation (3.27).

Ultimate State of Slope

As an alternative we study a possibility of a rupture in the plastic zone where elongations $\varepsilon_1 = \gamma/2$ take place. From expression (2.32) we write

$$\tau = 2G(t)\varepsilon_1 \exp(-\alpha\varepsilon_1)$$

and according to criterion $d\varepsilon_1/dt \to \infty$ we find the critical values of ε_1 and t as follows

$$\varepsilon_* = 1/\alpha, G(t_*) = p\alpha e(\cos 2(\lambda - \upsilon) - 1)/4(2\lambda \cos 2(\lambda - \upsilon) - 2\upsilon - \sin(\lambda - \upsilon)).$$

If the influence of time is negligible the ultimate load can be determined as

$$p_* = 4G(2\lambda \cos 2(\lambda - \upsilon) - 2\upsilon - \sin 2(\lambda - \upsilon))/\alpha e(\cos 2(\lambda - \upsilon) - 1).$$

The smallest value of p_* and p_u (see relation (4.11)) must be chosen.

4.2.2 Compression of Massif by Inclined Rigid Plates

Main Equations

Here we use the scheme in Fig. 3.6. Excluding from (2.66), (2.68) at $\tau_e = \tau_{yi}$ difference $\sigma_r - \sigma_\theta$ we get on an equation for $\tau_{r\theta} \equiv \tau$ at $\tau = \tau(\theta)$ which after the integration becomes

$$d\tau/d\theta = \pm\left(-2\sqrt{(\tau_{yi})^2 - \tau^2}\right) + 2n\tau_{yi} \qquad (4.12)$$

where n is a constant. The integration of (4.12) gives a row of useful results.

When $n = 0$ we find expression $\tau = \pm\tau_{yi} \sin(c + 2\theta)$ which corresponds to homogeneous tension or compression. The family of these straight lines has two limiting ones on which $\tau = \pm\tau_{yi}$ (they are called "slip lines") and according to the first two equations (3.21) $\sigma_r = \sigma_\theta = \pm 2\tau_{yi}\theta$. Another family of slip curves is a set of circular arcs (Fig. 4.4a), Such a field was realized in plastic zone BOC of the problem in Sect. 4.2.1 and can be seen near punch edges. The photographs of compressed marble and rock specimens are given in book /22/ and they are shown schematically in Fig. 4.4b. It is interesting

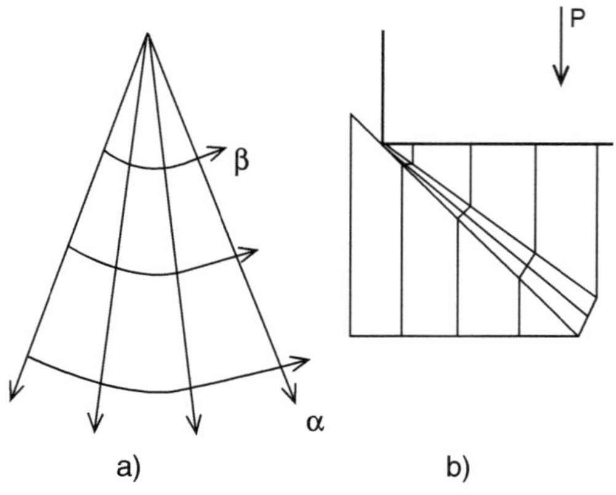

Fig. 4.4. Slip lines

to notice that this stress state is described by the same potential function (see (2.75))

$$\Phi = \tau_{yi} r^2 \theta$$

as in an elastic state.

General Case

When in (4.12) $n \neq 0$ we have a compression of a wedge by rough rigid plates. Putting in (4.12)

$$\tau = \tau_e \sin 2\psi, \sigma_r - \sigma_\theta = 2\tau_e \cos 2\psi \tag{4.13}$$

where ψ is equal to angle Ψ in Figs. 1.21 and 1.22 we find for the upper sign in (4.12)

$$d\psi/d\theta = n/\cos 2\psi - 1. \tag{4.14}$$

The integral of (4.14) at boundary condition $\psi(0) = 0$ is obvious

$$\theta = n(n^2 - 1)^{-1/2} \tan^{-1}\left(\sqrt{(n+1)/(n-1)} \tan \psi\right) - \psi$$

and n depends on λ according to the second border demand $\psi(\lambda) = \pi/4$ as (Fig. 4.5)

$$\lambda = n(n^2 - 1)^{-1/2} \tan^{-1} \sqrt{(n+1)/(n-1)} - \pi/4.$$

Now from static equations (2.67) we compute

$$\begin{matrix}\sigma_r \\ \sigma_\theta\end{matrix} = \tau_{yi}(C - 2\ln(r/a) - n\ln(n - \cos 2\psi) \pm \cos 2\psi) \tag{4.15}$$

where a is represented in Fig. 3.6 and constant C can be found from the first equation (3.32). The simplest option is

$$\begin{matrix}\sigma_r \\ \sigma_\theta\end{matrix} = \tau_{yi}(2\ln(a/r) - n\ln((n - \cos 2\psi)/(n-1)) \pm \cos 2\psi).$$

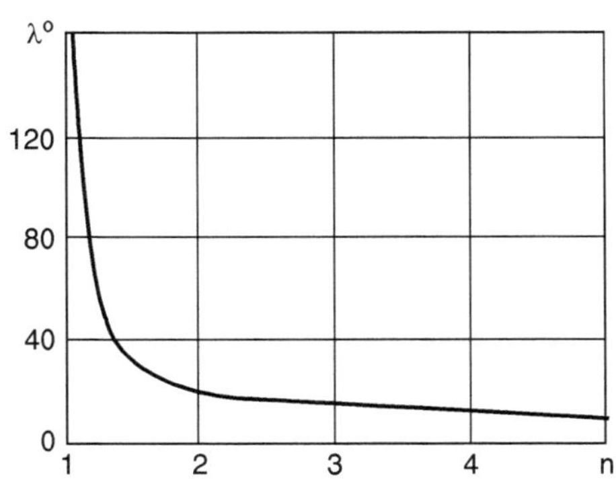

Fig. 4.5. Dependence λ on n

In order to find displacements we derive from (2.66), (3.18) and (4.13)

$$(U' - r^2 f(\theta))/U = -2\tan 2\psi.$$

Since $\psi = \psi(\theta)$ we must put $f(\theta) = V' + V = 0$ and deduce together with conditions $V(0) = 0, V(\lambda) = V_o$ and (4.14)

$$u_\theta = V_o \sin\theta/\sin\lambda, u_r = u_o/r(n - \cos 2\psi) - V_o\cos\theta/\sin\lambda$$

where u_o may be found from the supposition that some point at $\theta = \lambda$, $\psi = \pi/4, r = d$ has no radial movement. As a result we have

$$u_r = V_o(nd/r(n - \cos 2\psi)\tan\lambda - \cos\theta/\sin\lambda)$$

Sokolovski /18/ used this solution for the description of material flow through a narrowing channel. For both cases we can find resultant $Q = ql$ (Fig. 3.6) according to the second integral static equation (3.32) as /23/

$$q = 2n\tau_{yi}((a/l + 1)\ln(l/a + 1) + 0.5\ln(n/(n - 1)) - 1).$$

Diagram $\max\tau_e(\lambda)$ at $l/a \to \infty$ is given in Fig. 3.9 by solid curve 0. We can see that at $\lambda > \pi/4$ it is below the line 1.

Cases of Big n and Parallel Plates

If n is high we have from (4.14)

$$d\psi/d\theta = n/\cos 2\psi$$

and after integration

$$n\theta = 0.5\cos 2\psi$$

Parameter n is linked with λ as $n = 1/2\lambda$ and for ψ we have

$$\sin 2\psi = \theta/\lambda,\ \cos 2\psi = \sqrt{1 - (\theta/\lambda)^2}.$$

In the same manner as before we find stresses

$$\tau = \tau_{yi}\theta/\lambda,\ \frac{\sigma_r}{\sigma_\theta} = \tau_{yi}\left(\lambda^{-1}\ln(a/r) - 1 + \frac{2\sqrt{1 - (\theta/\lambda)^2}}{0}\right),$$

and as before we find displacements. From (2.66), (3.28), (4.13) we have

$$(U' - f(\theta)r)/U = 2\theta/\sqrt{\lambda^2 - \theta^2}.$$

That means $f(\theta) = 0$ and

$$\ln(U/u_o) = -2\sqrt{\lambda^2 - \theta^2}$$

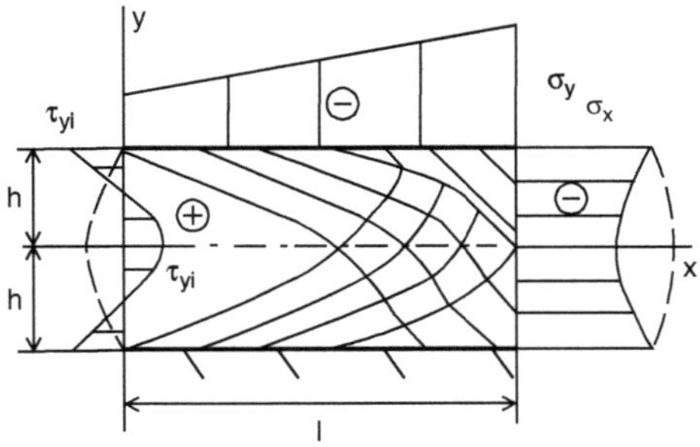

Fig. 4.6. Compression of massif by parallel plates

from which we derive

$$u_\theta = V_o\theta/\lambda,\ u_r = V_o\left((d/r)\exp\left(-\sqrt{\lambda^2-\theta^2}\right)-1\right)/\lambda.$$

Lastly at $\lambda \to 0$ we have the case of parallel plates and at $\theta/\lambda = y/h$ we derive (Fig. 4.6)

$$\tau = \tau_{yi}y/h,\ \sigma_x - \sigma_y = 2\tau_{yi}\sqrt{1-(y/h)^2}$$

and using (2.59) at $X = Y = 0$

$$\sigma_x = f(y) - \tau_{yi}x/h,\ f'(y) = -2\tau_{yi}y/h\sqrt{h^2-y^2}$$

or after integration

$$\sigma_x = (C - \tau_{yi}x/h) + 2\tau_{yi}\sqrt{1-(y/h)^2},\ \sigma_y = C - \tau_{yi}x/h.$$

Supposing $\sigma_y(0,h) = -\tau_{yi}$ we finally derive

$$\sigma_x = -\tau_{yi}\left(1 + x/h - 2\sqrt{1-(y/h)^2}\right),\ \sigma_y = \tau_{yi}(1 + x/h) \qquad (4.16)$$

and from integral static equation we compute

$$p = P/l = \tau_{yi}(1 + l/2h).$$

Diagrams $\sigma_x(y)$. $\sigma_x(l)$ and $\sigma_y(x)$ for the left side of the layer are shown in Fig. 4.6. The broken lines correspond to the case when the material is pressed into space between the plates (two similar states are described in Sect. 1.5.4).

In order to seek displacements we suppose $u_y = -V_o y/h$ and according to (2.60), (3.42) and incompressibility condition ($\varepsilon_x + \varepsilon_y = 0$) we have

$$u_x = F(y) + V_o x/h,\ \gamma_{xy} = F'(y).$$

and from the similarity of deviators

$$\gamma_{xy}/\varepsilon_x = 4\tau/(\sigma_x - \sigma_y)$$

with consideration of (2.66)

$$F'(y) = 2V_o y/h\sqrt{h^2 - y^2}$$

or after integration together with border demand $u_x(l, h) = 0$

$$u_x = V_o\left((x - l)/h - 2\sqrt{1 - (y/h)^2}\right).$$

The set of slip lines is also drawn in Fig. 4.6. They are cycloids and their equation will be given later. Experimental investigations show that rigid zones appear near the centre of the plate (shaded districts in Fig. 4.6) while plastic material is pressed out according to the solution (4.16) above.

Its analysis shows that at small h/l shearing stresses are much less than the normal ones and the material is in a state near to a triple equal tension or compression. This circumstance has big practical and theoretical meaning. It explains particularly the high strength of layers with low resistance to shear in tension (solder, glue etc.) or compression (soft material between hard one in nature or artificial structures). It also opens the way to applied theory of plasticity /10/.

Generalization of this problem on hardening at creep body is given in Appendix H.

Addition of Shearing Force

Here we suppose /10/ that shearing stresses on contact surfaces (Fig. 4.7) are constant. At $y = h, x < l$ and $y = -h$, $x > l$ we have $\tau = \tau_{yi}$ and in other parts of the surface $\tau = \tau_1 < \tau_{yi}$. Then satisfying static equations (2.59) and condition $\tau_e = \tau_{yi}$ the solution may be represented in a form

$$\begin{aligned} \tau_{xy}/\tau_{yi} &= (1 + k_1)/2 + (1 - k_1)y/2h, \\ \sigma_y/\tau_{yi} &= -C - (1 - k_1)x/2h,\ \sigma_x/\tau_{yi} = \sigma_y/\tau_{yi} + 2\sqrt{1 - (\tau_{xy}/\tau_{yi})^2}. \end{aligned} \quad (4.17)$$

Here $k_1 = \tau_1/\tau_{yi}$ and C is a constant. If $k_1 = -1$ we have solution (4.16) and at $k_1 = 1$ we receive a pure shear ($\sigma_x = \sigma_y = 0, \tau_{xy} = \tau_{yi}$).

Now we use integral static equations similar to (3.32)

$$\int_{-h}^{h} \sigma_x(0, y)dy = 0,\ \int_{0}^{l} \sigma_y(h)dx = P$$

which give after exclusion of C

$$\pi/2 - k_1\sqrt{1 - (k_1)^2} - \sin^{-1} k_1 = (1 - k_1)(-p - (1 - k_1)l/4h). \quad (4.18)$$

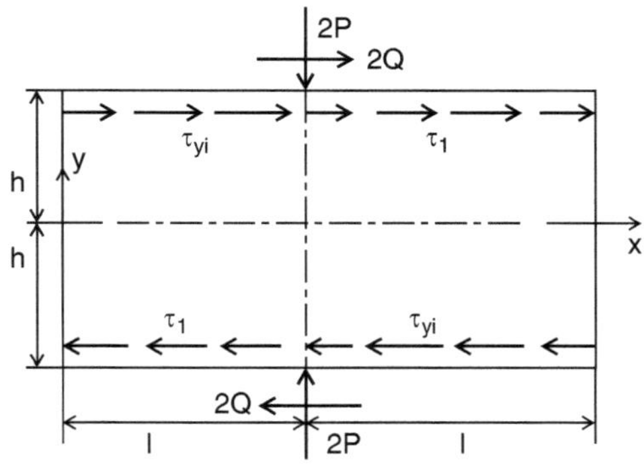

Fig. 4.7. Layer under compression and shear

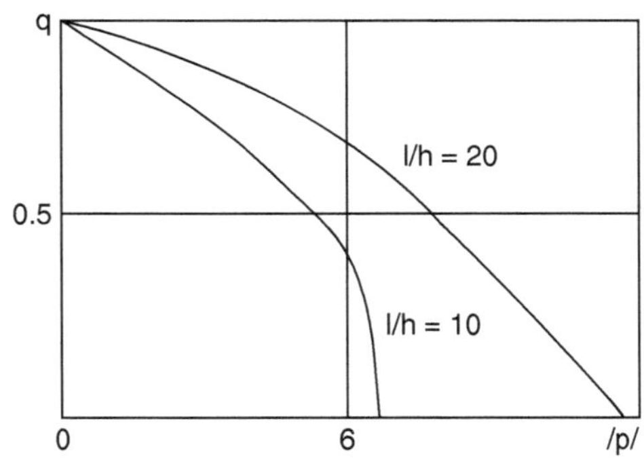

Fig. 4.8. Dependence of p on q

Here $p = P/l\tau_{yi}$. Then we take integral equilibrium equation at contact surface as

$$2Q = \tau_{yi}(1 + k_1)l$$

which gives $1 + k_1 = 2q$ where $q = Q/\tau_{yi}l$. Excluding from (4.18) k_1 we finally receive

$$(1 - q)(-2p - (1 - ql/h) = \pi/2 + 2(1 - 2q)\sqrt{q(1 - q)} - \sin^{-1}(2q - 1). \quad (4.19)$$

At $q = 0$ we again find the Prandtl's solution (4.16).

From Fig. 4.8 where diagrams (4.19) for $l/h = 10$ and $l/h = 20$ are constructed we can see the high influence of q on ultimate pressure p.

4.2.3 Penetration of Wedge and Load-bearing Capacity of Piles Sheet

As we can see from Fig. 4.5 the dependence $\lambda(n)$ may be also used at $\lambda > \pi/2$ when a wedge penetrates into a medium (Fig. 4.9).General relations for

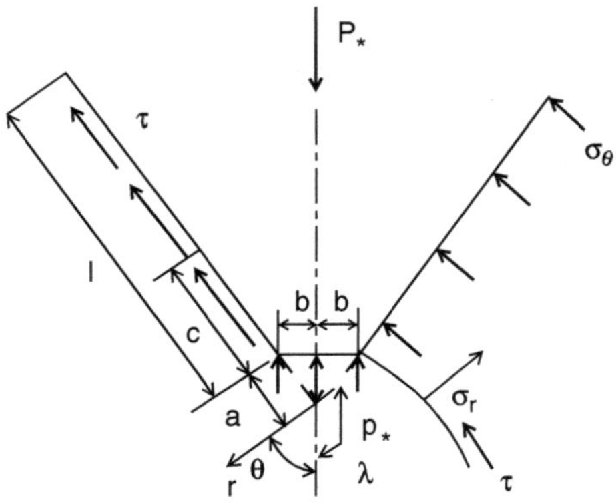

Fig. 4.9. Penetration of wedge

stresses of Sect. 4.2.2 are valid here but constant C should be searched from equations similar to (3.32) as

$$p_* \sin\lambda = -\int_0^{\lambda} (\sigma_r(a,\theta)\cos\theta + \tau(a,\theta)\sin\theta)d\theta,$$

$$P_* = 2(p_* b + \int_a^{a+l} (\sigma_\theta(r,\lambda)\sin\lambda + \tau(r,\lambda)\cos\lambda)dr). \qquad (4.20)$$

where p_* is an ultimate pressure at compression. Putting into (4.20) σ_r, σ_θ from (4.15) and τ from (4.13) we find

$$P_*/2\tau_{yi} = p_*(b/l + \sin\lambda)/\tau_{yi} + J_o - n(\ln n - 2 + 2n(1 + a/l) \times \ln(l/a + 1))\sin\lambda + \cos\lambda. \qquad (4.21)$$

Here

$$J_o = \int_0^{\lambda} (\cos 2\psi - n\ln(n - \cos 2\psi)\cos\theta + \sin 2\psi \sin\theta)d\theta.$$

In the case of a wedge penetration we must put in (4.21) $a = 0$ that gives the infinite ultimate load due to the hypothesis of constant form and volume of the material near the wedge. Because of that we recommend for the case the solution of Sect. 4.2.2. However for λ near π (an option of pile sheet) simple engineering relation can be derived when at $n = 1.07$, $\lambda = 179°, a \to \infty$ we derive from (4.21)

$$P_* = 2(p_* b + \tau_{yi} l(1 + J_o)). \qquad (4.22)$$

The computations of $J_o(\pi)$ gives its value 2.0. Taking into account the structure of (4.22) and its original form (4.20) we can conclude that the influence of σ_θ is somewhat higher than that of τ. We must also notice that P_*-value in (4.22) is computed in the safety side because we do not consider an influence of σ_θ on τ_{yi}.

4.2.4 Theory of Slip Lines

Main Equations

Such rigorous results as in previous paragraphs are rare. More often approximate solutions are derived according to the theory of slip lines that can be observed on polished metal surfaces. They form two families of perpendicular to each other lines for materials with τ_{yi} = constant. We denote them as α, β and for their determination we use transformation relations (2.72) which give the following stresses in directions inclined to main axes 1, 3 under angles $\pi/4$ (Fig. 4.10)

$$\sigma_\alpha = \sigma_\beta = \sigma_m = 0.5(\sigma_1 + \sigma_3),$$
$$\tau_{\alpha\beta} = \tau_{yi} = 0.5(\sigma_1 - \sigma_3). \tag{4.23}$$

Now we find the stresses for a slip element in axes x, y. According to expressions (2.72) (Fig. 4.11)

$$\begin{matrix}\sigma_x \\ \sigma_y\end{matrix} = \sigma_m \pm \tau_{yi} \sin 2\psi,\ \tau_{xy} = -\tau_{yi} \cos 2\psi. \tag{4.24}$$

These relations allow to find equations of slip lines in form

$$dy/dx = \tan\psi = (1 - \cos 2\psi)/\sin 2\psi = 2(\tau_{yi} + \tau_{xy})/(\sigma_x - \sigma_y)$$

and for another family $dy/dx = -\cot\psi$.

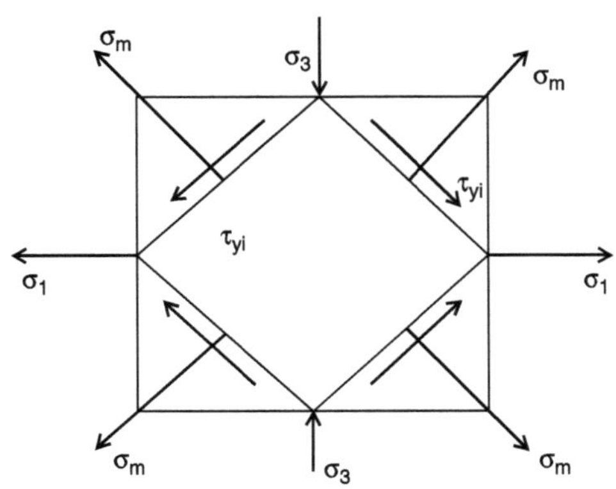

Fig. 4.10. Stresses in element at ideal plasticity

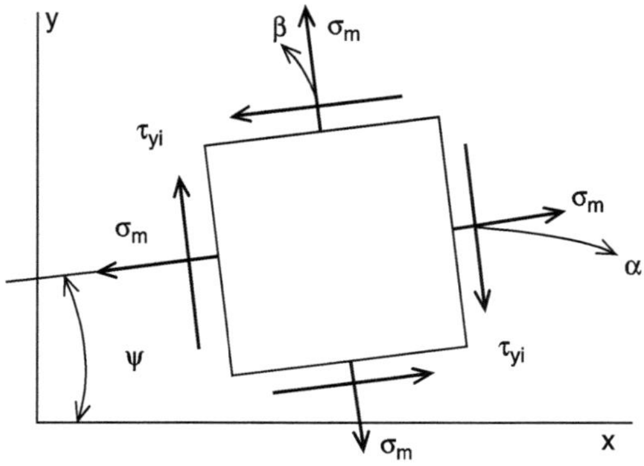

Fig. 4.11. Slip element in axes x, y

Examples of Slip Lines

Reminding the problem of the layer compression (see Sect. 4.2.2.) we put in the last expressions the relations for stresses and get on equations

$$\mathrm{dy/dx} = -\sqrt{(\mathrm{h}-\mathrm{y})/(\mathrm{h}+\mathrm{y})},\ \mathrm{dy/dx} = \sqrt{(\mathrm{h}+\mathrm{y})/(\mathrm{h}-\mathrm{y})}$$

and after integration we find the both families of the slip lines as

$$\mathrm{x} = \mathrm{C} + \sqrt{\mathrm{h}^2-\mathrm{y}^2} + \mathrm{h}\cos^{-1}(\mathrm{y/h}),\ \mathrm{x} = \mathrm{C} + \sqrt{\mathrm{h}^2-\mathrm{y}^2} - \mathrm{h}\cos^{-1}(\mathrm{y/h})$$

where C is a constant. The slip lines according to these expressions are shown in Fig. 4.6. In a similar way the construction of slip lines can be made for the compressed wedge in Fig. 3.6.

As the second example we consider a tube with internal a and external b radii under internal pressure q. Here $\tau_{r\theta} = 0, \sigma_r - \sigma_\theta = 2\tau_e = \sigma_{yi}$ and from the first static equation (2.67) we receive

$$q_* = \sigma_{yi}\ln(b/a). \tag{4.25}$$

Slip lines are inclined to axes r and θ by angle $\pi/4$ (broken lines in Fig. 4.12). From this figure we also find differential equation

$$dr/rd\theta = \pm 1$$

with an obvious integral

$$r = r_o \exp(\pm\theta). \tag{4.26}$$

So, the slip lines are logarithmic spirals which can be seen at pressing of a sphere into a plastic material.

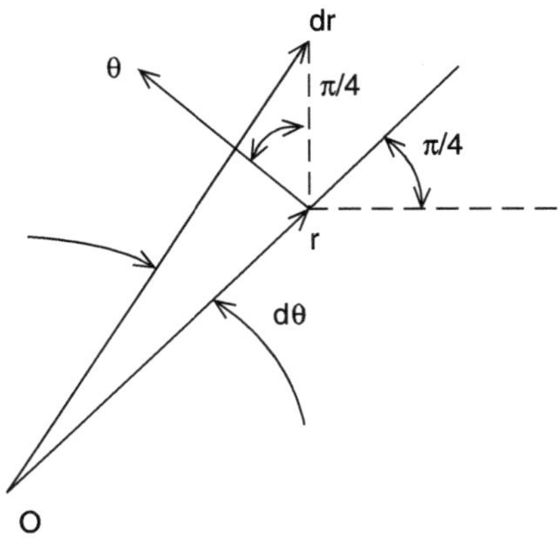

Fig. 4.12. Slip lines in tube under internal pressure

Construction of Slip Lines Fields

In order to construct a more general theory of slip lines we transform static equations (2.59) into coordinates α, β putting there expressions (4.24). Applying the method of Sect. 2.4.3 (see also /10/) we derive differential equations

$$\partial(\sigma_m + 2\tau_{yi}\psi)/\partial\alpha = 0, \partial(\sigma_m + 2\tau_{yi}\psi)/\partial\beta = 0$$

with obvious integrals

$$\sigma_m/2\tau_{yi} \pm \psi = \frac{\xi}{\eta} = \text{constant.} \tag{4.27}$$

The latter relations allow to determine parameters ξ, η in a whole field if they are known on some its parts particularly on borders. In practice simple constructions are used corresponding as a rule to axial tension or compression (Fig. 4.13) and centroid one (Fig. 4.4a). A choice between different options should be made according to the Gvozdev's theorems /9/.

Construction of Slip Fields for Soils

In a similar way the simple fields of slip lines can be found for a soil with angle of internal friction φ (see solid straight line in Fig. 1.22) when according to (1.34), (1.35) the slip planes in a homogeneous stress field are inclined to the planes with maximum and minimum main stresses under angles $\pi/4-\varphi/2$ and $\pi/4+\varphi/2$ respectively. In order to generalize the centroidal field in Fig. 4.4a we find from Fig. 1.22 expression $\tau = \pm(-\sigma_\theta \tan\varphi)$ and put it into the second equation (3.21) which after transformations gives

$$\sigma_\theta = C\exp(\pm 2\theta\tan\varphi), \tau = \pm(-C(\tan\varphi)\exp(\pm 2\theta\tan\varphi)).$$

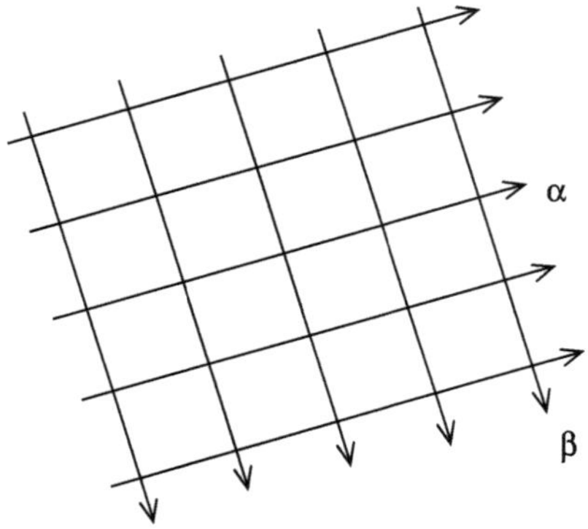

Fig. 4.13. Slip lines at homogeneous tension or compression

Now we again use Fig. 1.22 and write the result at the upper sign in the previous relations as follows

$$\sigma_m = \sigma_\theta + \tau \tan\varphi = C(1 + \tan^2\varphi)\exp(\pm 2\theta \tan\varphi)$$

or finally

$$\sigma_m = D\exp(\pm 2\theta\tan\varphi) \tag{4.28}$$

where D is a constant.

Supposing that in the origin at $r = r_o$ the second family of the slip lines is inclined to the first set of them (the rays starting from the centre – see Fig. 4.4) under angle $\pi/4 - \varphi/2$ we conclude from Fig. 1.22 that they form angle φ with the normal to r. So for the second family we have equation similar to the case of τ_{yi} = constant as

$$dr/rd\theta = \pm\tan\varphi$$

and hence

$$r = r_o\exp(\pm\theta\tan\varphi) \tag{4.29}$$

(see also (4.26) and Fig. 4.12). This theory can be generalized for a cohesive soil by the replacement in (4.28) σ_m by $\sigma_m + c/\tan\varphi$ (broken line in Fig. 1.22).

4.2.5 Ultimate State of Some Plastic Bodies

Plate with Circular Hole at Tension or Compression

We begin with a simple example of a circular tunnel (Fig. 4.14) in a massif under external homogeneous pressure p. In this case we choose a slip lines field corresponding to simple compression (left side in the figure). Then we have according to relations (4.27) $\sigma_x = \sigma_m + \tau_{yi} = 0$ that means $\sigma_m = -\tau_{yi}$ and $\sigma_y = \sigma_m - \tau_{yi} = -2\tau_{yi} = -\sigma_{yi}$. We suppose also that the material inside a strip 2a is rigid and we find

$$P_* = 2(b - a)\sigma_{yi}. \tag{4.30}$$

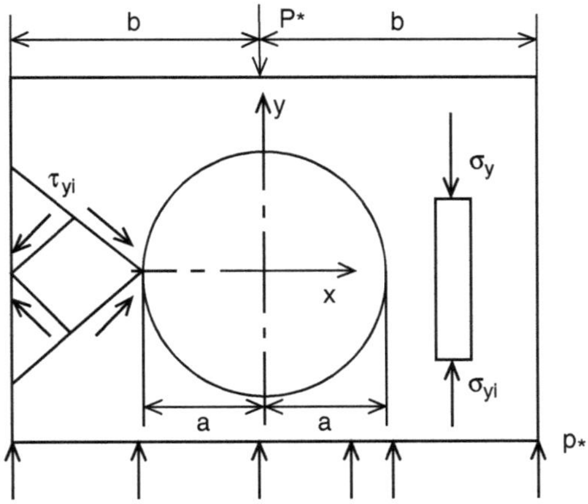

Fig. 4.14. Compression of massif with circular tunnel

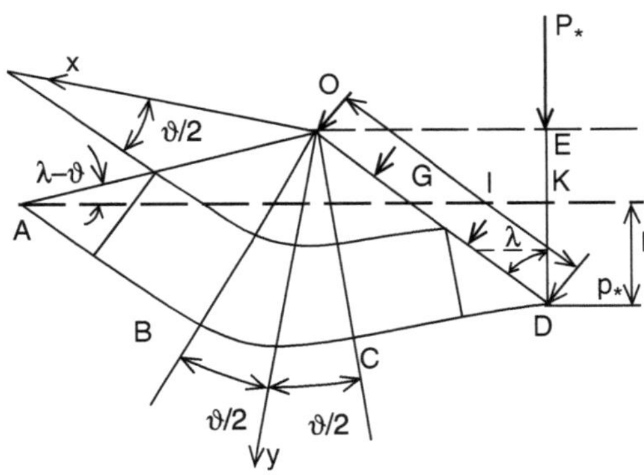

Fig. 4.15. Penetration of wedge

As the P_*-value is found from the static equation the result is a rigorous one. It is also valid for a tension of the plane with a circular hole and it is much simpler than the similar solution for an elastic body in Sect. 3.2.5.

Penetration of Wedge

Now we consider a pressure of a wedge into a massif (Fig. 4.15). We suppose that a new surface OA is a plane and the slip field consists of two triangles OAB, OCD at pure compression and a centroidal part OBC between them. Firstly we determine the stress state in the triangles. In AOB

$$\psi = -\upsilon/2, \quad \sigma_1 = \sigma_m + \tau_{yi} = 0$$

that means $\sigma_m = -\tau_{yi}$. Similarly in COD

$$\psi = \upsilon/2, \sigma_3 = -p_* = \sigma_m - \tau_{yi}$$

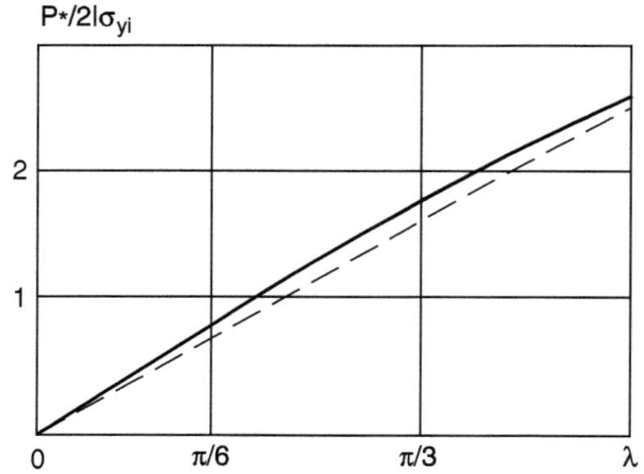

Fig. 4.16. Dependence of compressing force on angle λ

that gives $\sigma_m = \tau_{yi} - p_*$.

Putting these results into (4.27) we receive

$$-\tau_{yi}/2\tau_{yi} - \upsilon/2 = (\tau_{yi} - p_*)/2\tau_{yi} + \upsilon/2$$

from which

$$p_* = \sigma_{yi}(1 + \upsilon) \tag{4.31}$$

and according to static equation as the sum of the forces on vertical direction:

$$P_* = 2\sigma_{yi}(1 + \upsilon)l \sin\lambda. \tag{4.32}$$

The auxiliary quantity υ can be excluded by the condition of the equality of volumes KDG and AOG. Since from triangle AOG angle OAG is equal to $\pi - (\pi/2 - \lambda) - (\pi/2 + \upsilon)$ or after cancellation – to $\lambda - \upsilon$ we have for segment KE

$$l\cos\lambda - h = l\sin(\lambda - \upsilon) \tag{4.33}$$

and we find /24/

$$h^2 \tan\lambda = (l\cos\lambda - h)(l\cos(\lambda - \upsilon) + (l\cos\lambda - h)\tan\lambda). \tag{4.34}$$

Excluding from (4.33), (4.34) l, h we finally derive

$$2\lambda = \upsilon + \cos^{-1}(\tan(\pi/4 - \upsilon/2)). \tag{4.35}$$

Diagram $P_*(\lambda)$ according to (4.32), (4.35) is represented in Fig. 4.16 by broken line. Replacing in (4.31) υ by $2\lambda - \pi/2$ we get on the critical pressure (4.11) for the slope.

Pressure of Massif through Narrowing Channel

Similar to investigations of the previous paragraph we can study the scheme in Fig. 4.17. We consider first the option $l = h$ and the slip lines field consisting

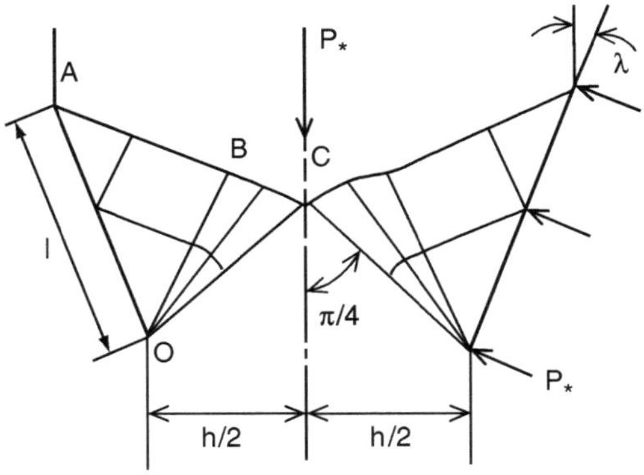

Fig. 4.17. Pushing massif through channel

of triangle AOB and sector OBC on each half. The parameters in the triangle and on straight line OC are respectively

$$\psi = \lambda + \pi/4, \sigma_3 = \sigma_m - \tau_{yi} = -p_*; \psi = \pi/4, \sigma_1 = \sigma_m + \tau_{yi} = 0. \tag{4.36}$$

Putting (4.36) into (4.27) we have

$$p_* = 2\tau_{yi}(1 + \lambda) \tag{4.37}$$

and from static equation we finally receive

$$P_* = 2l\sigma_{yi}(1 + \lambda)\sin\lambda. \tag{4.38}$$

Relation (4.38) is represented in (Fig. 4.16) by the solid line and we can see that it is near to the broken curve which corresponds to the latter solution for $b = 0$. So, we can conclude that the simple results (4.37), (4.38) can be used for a case of $l > h$ as well.

At $\upsilon = \pi/2$ in (4.31) and $\lambda = \pi/2$ in (4.37) we find the ultimate punch pressure (left part in Fig. 4.18) as

$$p_* = \sigma_{yi}(1 + \pi/2). \tag{4.39}$$

Tension of Plane with Crack

Relation (4.39) is valid for the problem of a crack in tension (right part in Fig. 4.18). Here in square ODCD'

$$\sigma_x = \sigma_{yi}\pi/2, \tau_{xy} = 0, \sigma_y = \sigma_{yi}(1 + \pi/2)$$

and according to (2.72) we compute

$$\sigma_r = \sigma_{yi}(\pi/2 + \sin^2\theta), \sigma_\theta = \sigma_{yi}(\pi/2 + \cos^2\theta), \tau_{r\theta} = \tau_{yi}\sin 2\theta. \tag{4.40}$$

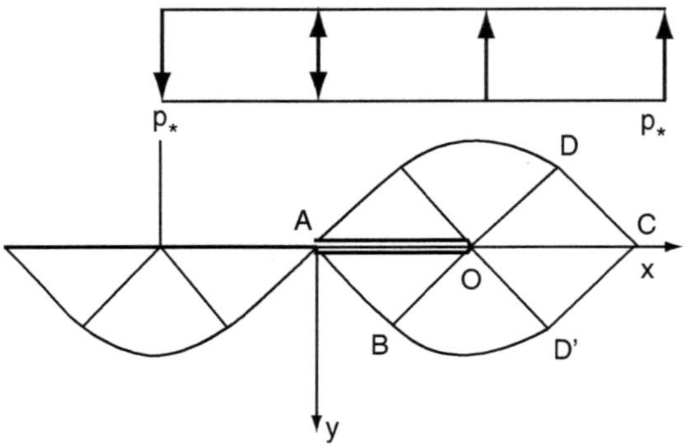

Fig. 4.18. Pressure of punch and tension of plate with crack

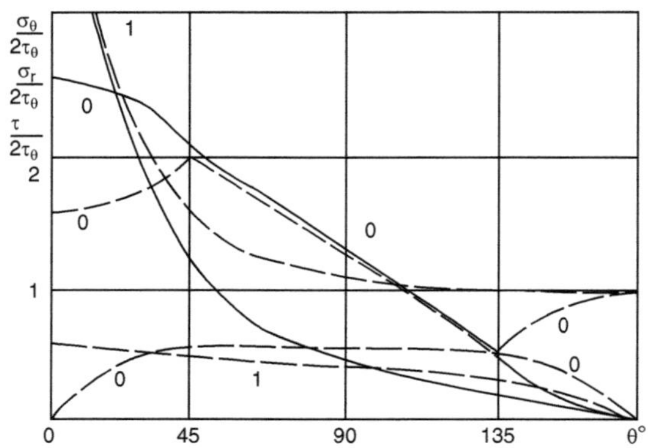

Fig. 4.19. Diagrams of stresses

In the same manner we find in triangle AOB $\sigma_y = \tau_{xy} = 0$, $\sigma_x = \sigma_{yi}$ and

$$\sigma_r = \sigma_{yi}\cos^2\theta, \sigma_\theta = \sigma_{yi}\sin^2\theta, \tau_{r\theta} = -\tau_{yi}\sin 2\theta. \tag{4.41}$$

In sector OBD' $\tau_{r\theta} = \tau_{yi}$ and stresses $\sigma_r = \sigma_\theta$ change as linear function of θ:

$$\sigma_r = \sigma_\theta = \sigma_{yi}(0.5 + 3\pi/4 - \theta). \tag{4.42}$$

Diagrams $\sigma_\theta/\sigma_{yi}$, σ_r/σ_{yi}, $\tau_{r\theta}/\sigma_{yi}$ are represented in Fig. 4.19 by solid, broken and interrupted by points lines 0. The same curves with index 1 refer to elastic solution (3.90), at $\max\tau_e = \sigma_{yi}/2$. It is interesting to notice that these lines reflected relatively axis $\theta = \pi/2$ describe the stress state near the punch edge.

It is also valid to note that in plastic state the potential function exists near the crack ends as

$$0.5\,\sigma_{yi}r^2(\pi/2 + \cos^2\theta), 0.5\,\sigma_{yi}r^2(0.5 + 3\pi/4 - \theta), 0.5\,r^2\sigma_{yi}\sin^2\theta \tag{4.43}$$

at $\theta \le \pi/4$, $\pi/4 \le \theta \le 3\pi/4$ and $3\pi/4 \le \theta \le \pi$ respectively.

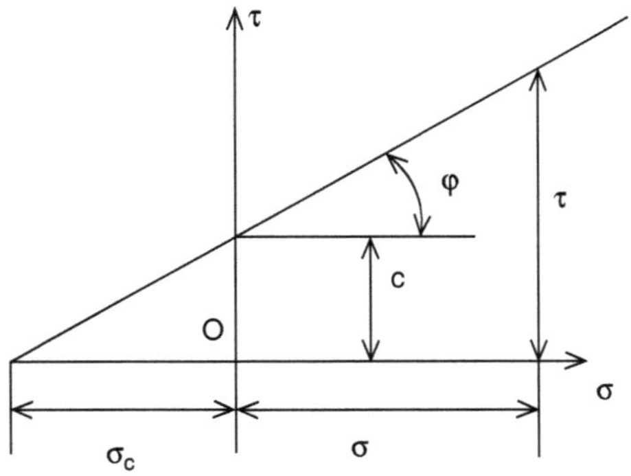

Fig. 4.20. Generalized Coulomb's law

4.2.6 Ultimate State of Some Soil Structures

Conditions of Beginning of Plastic Shear

As we noticed above an earth is a very complex medium and its fracture is usually linked with shearing stresses. The strength condition is as a rule written in form $\tau < \tau_*$ – stable equilibrium, $\tau = \tau_*$ – ultimate state and $\tau > \tau_*$ – plastic flow where τ_* is a characteristic of a material a value of which depends linearly on normal stress applied to the plane where τ acts. This is the Coulomb's law (here up to sub-chapter 4.3 according to /10/ compressive stresses are supposed positive with $\sigma_3 > \sigma_1$).

$$\tau_* = \sigma \tan \varphi \tag{4.44}$$

(inclined straight line in Fig. 1.22) for a quicksand and

$$\tau_* = \sigma \tan \varphi + c \tag{4.45}$$

(inclined broken line in the figure) – for a coherent soil. The latter equality is usually led to the form (4.44) (Fig. 4.20)

$$\tau_* = (\sigma + \sigma_c) \tan \varphi \tag{4.46}$$

where $\sigma_c = c/\tan\varphi$ – coherent pressure which replaces an action of all cohesive forces.

From (4.46) we have

$$\tan \varphi = \tau_*/(\sigma + \sigma_c). \tag{4.47}$$

This condition may be written in another form. We draw through a point A (Fig. 4.21) at angle β to the horizon plane mn on which the components of full stress p – normal σ_β and shearing τ_β are acting. The first of them includes the cohesion pressure. From geometrical consideration we find

$$\tan \theta = \tau_\beta/(\sigma_\beta + \sigma_c). \tag{4.48}$$

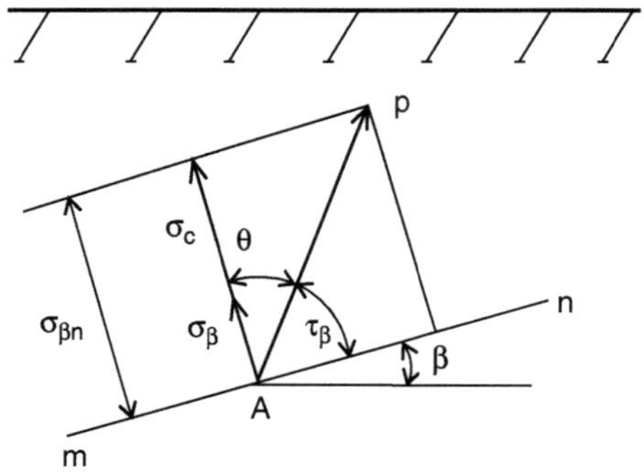

Fig. 4.21. Decomposition of full stress

Value of θ is usually called an angle of divergence which can not exceed angle of internal friction φ. That gives the condition of ultimate equilibrium as

$$\theta = \varphi. \quad (4.49)$$

Representations of Ultimate Equilibrium Condition

At an appreciation of materials' strength the so-called Mohr's circles are used. In the common representation of a tensor as a vector in a nine-dimensional space /10/ there are three such figures. At a plane stress state we have in coordinates σ, τ only one circumference (Fig. 1.22) along which a point moves when a plane turns in a material.

As was told in Chap. 2 the faces of a cube with absent shearing stresses are called main (principal) planes with normal stresses on them $\sigma_1 = \sigma_x$, $\sigma_2 = \sigma_z$, $\sigma_3 = \sigma_y$. O. Mohr used his representation for a formulation of his hypothesis of strength which in its linear option coincides with the Coulomb's relation (4.44) and can be interpreted as a tangent to the circumference in Fig. 1.22 under angle φ.

From expression (1.36) we have in main stresses the condition of the ultimate state of quicksand as:

$$\sin\varphi = (\sigma_3 - \sigma_1)/(\sigma_3 + \sigma_1). \quad (4.50)$$

For coherent earth (4.50) can be generalized in form (broken line in Fig. 1.22)

$$\sin\varphi = (\sigma_3 - \sigma_1)/(\sigma_1 + \sigma_3 + 2c\cot\varphi). \quad (4.51)$$

Relation (4.50) can be also represented in form:

$$\sigma_1/\sigma_3 = \tan^2(\pi/4 \pm \varphi/2). \quad (4.52)$$

In the theory of interaction of structures with an earth sign minus corresponds to active pressure of soil and plus – to its resistance. In quicksand or coherent earth shearing displacements occur on planes under angles $\pi/4 - \varphi/2$ to the direction of σ_3.

In some cases it is useful to write (4.50), (4.51) in stresses σ_x, σ_y, τ_{xy} with the help of (2.65) as follows:

$$\sin^2 \varphi = ((\sigma_y - \sigma_x)^2 + 4(\tau_{xy})^2)/(\sigma_y + \sigma_x)^2 \qquad (4.53)$$

for quicksand and

$$\sin^2 \varphi = ((\sigma_y - \sigma_x)^2 + 4(\tau_{xy})^2)/(\sigma_x + \sigma_y + 2c \cot \varphi) \qquad (4.54)$$

for coherent soils.

Wedge Pressed in Soil

We construct the field of slip lines as in Fig. 4.22 /25/ and we again suppose that OA is a straight line. From the figure we compute that it is inclined to horizon AK by angle $\lambda - \upsilon$ as in Fig. 4.16. From geometrical considerations we have $l = al_1$ where

$$a = (1 - \sin \varphi)(\exp(-\upsilon \tan \varphi))/\cos \varphi$$

and

$$h = l_1(a \cos \lambda - \sin(\lambda - \upsilon)). \qquad (4.55)$$

Putting (4.55) into the condition of constant volume similar to that for ideal plastic material we find expression:

$$h^2 \tan \lambda = (l_1)^2 \sin(\lambda - \upsilon)(\cos(\lambda - \upsilon) + \sin(\lambda - \upsilon) \tan \lambda)$$

which gives after transformations relation for $\tan \lambda$:

$$(4a \cos \upsilon + \sin 2\upsilon) \tan^2 \lambda - 2(a^2 - \cos 2\upsilon + 2a \sin \upsilon) \tan \lambda - \sin 2\upsilon = 0. \qquad (4.56)$$

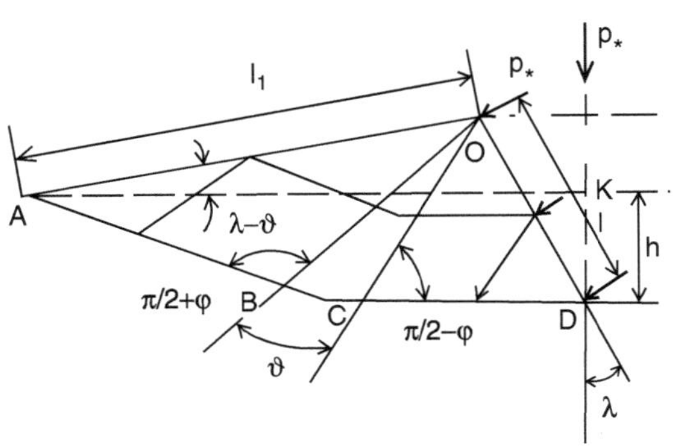

Fig. 4.22. Wedge pressed in soil

Now we find the ultimate load according to the field of slip lines in Fig. 4.22. From Fig. 1.22 we have for a cohesive soil:

$$\begin{matrix}\sigma_3\\ \sigma_1\end{matrix} = \sigma_m (1 \pm \sin\varphi) \pm c\cos\varphi. \tag{4.57}$$

In triangle ABO $\sigma_1 = \theta = 0$ and from (4.57)

$$\sigma_m(1 - \sin\varphi) = c\cos\varphi$$

but from (4.28) for a cohesive soil $\sigma_m = D - c/\tan\varphi$ and so

$$D = c/(1 - \sin\varphi)\tan\varphi. \tag{4.58}$$

In the same manner for triangle OCD where $\sigma_3 = p_*$, $\theta = \upsilon$ we find from (4.57), (4.28)

$$p_* = D(1 + \sin\varphi)(\exp 2\upsilon\tan\varphi) - c/\tan\varphi$$

and with consideration of D-value from (4.58) we receive finally:

$$p_* = c((1 + \sin\varphi)e^{2\upsilon\tan\varphi}/(1 - \sin\varphi) - 1)/\tan\varphi. \tag{4.59}$$

Lastly from static condition we derive:

$$P_* = 2lc((1 + \sin\varphi)e^{2\upsilon\tan\varphi}/(1 - \sin\varphi) - 1)\sin\lambda/\tan\varphi. \tag{4.60}$$

From diagrams $P_*/2lc = f(\lambda)$ at different φ in Fig. 4.23 we (broken curves) can see that P_* increases with a growth of φ and λ. It can be much bigger its value at ideal plasticity ($\varphi = 0, c = \tau_{yi}$ – broken line in Fig. 4.16).

Pressure of Soil through Narrowing Channel

Similar to Fig. 4.17 we can construct the field of slip lines for the soil (its right part is shown in Fig. 4.24). Then we use relations (4.57)–(4.60) with replacement in them υ by λ. The consequent diagrams are represented by solid lines in Fig. 4.23 and we can see that they are near to broken curves for more complex solution above.

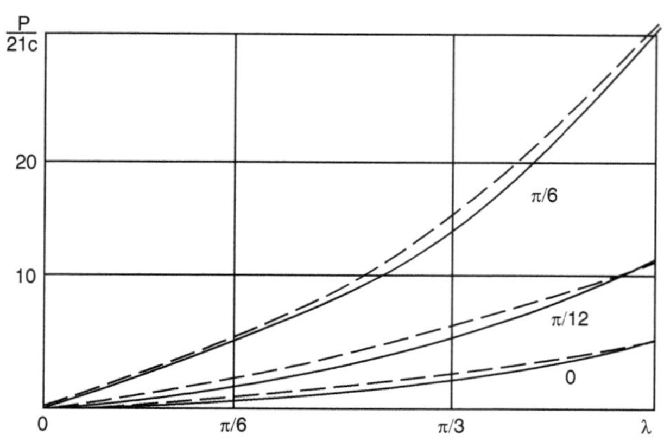

Fig. 4.23. Dependence of P_* on λ

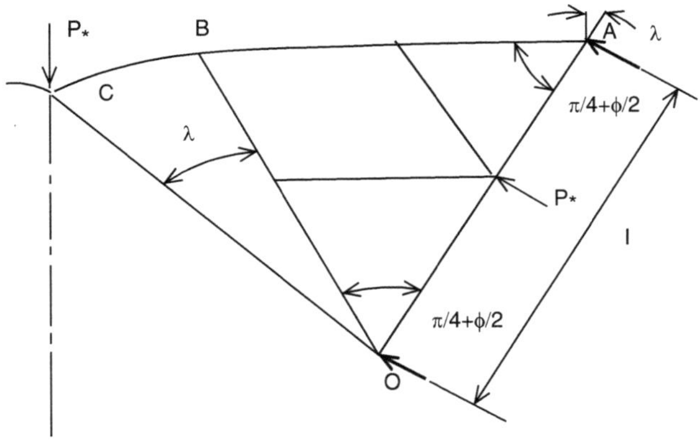

Fig. 4.24. Pressure of soil through narrowing channel

Some Important Particular Cases

At $\upsilon = \pi/2 - \beta$ we have from (4.59) the ultimate load for a slope (Fig. 3.5) as follows:

$$p_* = c((1 + \sin\varphi)e^{(\pi-2\beta)\tan\varphi}/(1 - \sin\varphi) - 1)\cot\varphi \tag{4.61}$$

and if $\upsilon = \pi/2$ – well-known p_u-value for a foundation (Fig. 3.12) – the so-called second ultimate load as:

$$p_* = (\gamma_e h + c\cot\varphi)(1 + \sin\varphi)e^{\pi\tan\varphi}/(1 - \sin\varphi) - c\cot\varphi. \tag{4.62}$$

At $h = 0$ we again see that this expression can not be applied to soils without cohesiveness. Once more we suppose $p_* = a_2\tau_{yi}$. The values of a_2 are given in the Table in Sect. 3.2.2. The ratio a_1/a_2 there means $p_*/\min p_{yi}$ and we can see that it increases strongly with a growth of φ.

4.2.7 Pressure of Soils on Retaining Walls

Active Pressure of Soil's Self-weight

A horizontal plane behind a vertical wall endures compression stress:

$$\sigma_3 = \gamma_e z. \tag{4.63}$$

Using equation of ultimate state (4.52) we find:

$$\sigma_1 = \gamma_e z\tan^2(\pi/4 - \varphi/2). \tag{4.64}$$

Diagram $\sigma_1(z)$ is given in Fig. 4.25 as triangle abd. The resultant of this pressure can be derived in form

$$R_a = 0.5\gamma_e H^2\tan^2(\pi/4 - \varphi/2). \tag{4.65}$$

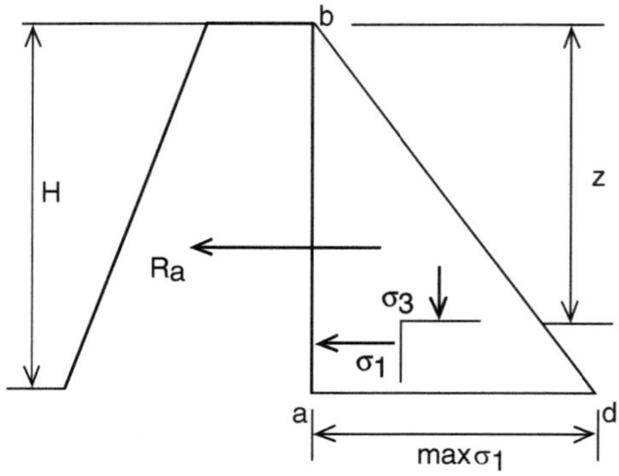

Fig. 4.25. Pressure of soil on vertical retaining wall

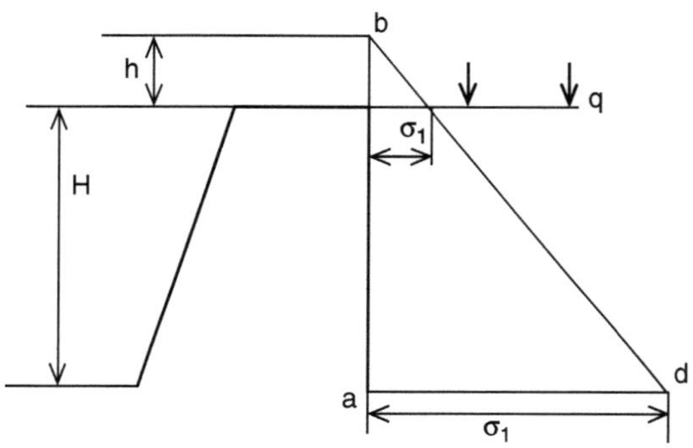

Fig. 4.26. Additional external pressure

In the case of the earth's passive resistance we must take in brackets of expressions (4.64), (4.65) sign plus.

When an uniformly distributed load q acts on a horizontal surface $z = 0$ we usually replace it by equivalent height $h = q/\gamma_e$ (Fig. 4.26) and the resultant is:

$$R = 0.5(\sigma_1 + (\sigma_1)')H.$$

Since

$$\sigma_1 = \gamma_e(H + h)\tan^2(\pi/4 - \varphi/2), (\sigma_1)' = \gamma_e h \tan^2(\pi/4 - \varphi/2) \tag{4.66}$$

the resultant can be computed as

$$R = 0.5\gamma_e H(H + 2h)\tan^2(\pi/4 - \varphi/2). \tag{4.67}$$

Consideration of Coherence

If a soil has coherence its influence can be conditionally replaced by three-dimensional pressure of coherence $\sigma_c = c/\tan\varphi$ (Fig. 1.22) and by equivalent layer

$$\sigma_c/\gamma_e = c/\gamma_e \tan\varphi. \tag{4.68}$$

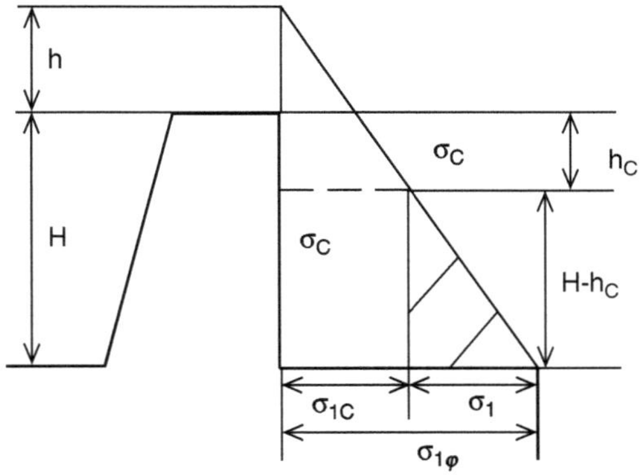

Fig. 4.27. Consideration of coherence

Taking this into account we can write:

$$\sigma_1 = \gamma_e(H + c/\gamma_e \tan\varphi)\tan^2(\pi/4 - \varphi/2) - c/\tan\varphi \text{ or}$$
$$\sigma_1 = \gamma_e H \tan^2(\pi/4 - \varphi/2) - 2c\tan(\pi/4 - \varphi/2). \tag{4.69}$$

According to Fig. 4.27 we can represent (4.69) as:

$$\sigma_1 = \sigma_{1\varphi} - \sigma_{1c} \tag{4.70}$$

where $\sigma_{1\varphi}$, σ_{1c} are maximum lateral pressures in an absence of the coherence and decrease of it due to coherent forces.

The whole pressure σ_1 changes from tension in the top to compression in the bottom and condition $\sigma_1 = 0$ gives:

$$h_c = 2c/\gamma_e \tan(\pi/4 - \varphi/2). \tag{4.71}$$

The resultant of active pressure can be found as the area of shaded triangle with base σ_1 and height $H - h_c$ that is:

$$R_c = 0.5\,\sigma_1(H - h_c).$$

Putting here σ_1 according to (4.69) we compute:

$$R_c = 0.5\gamma_e H^2 \tan^2(\pi/4 - \varphi/2) - 2cH\tan(\pi/4 - \varphi/2) + 2c^2/\gamma_e. \tag{4.72}$$

Comparing this result to (4.65) we can conclude that the coherence may diminish a resultant very strongly.

4.2.8 Stability of Footings

Besides the failures considered above a structure may loose its stability. We consider two types of such a phenomenon – plane and deep shears (Figs. 4.28 and 4.29 respectively).

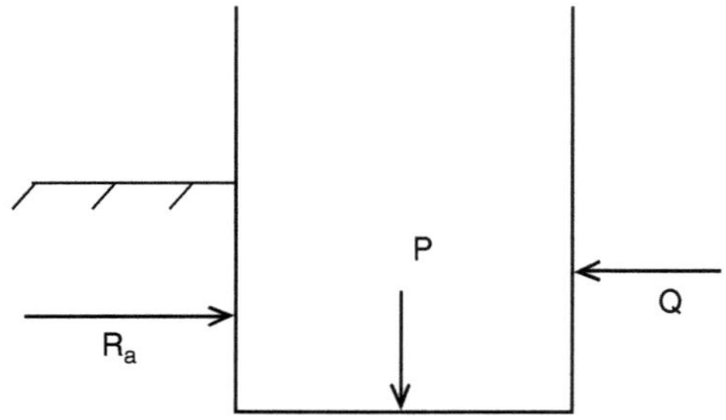

Fig. 4.28. Plane shear

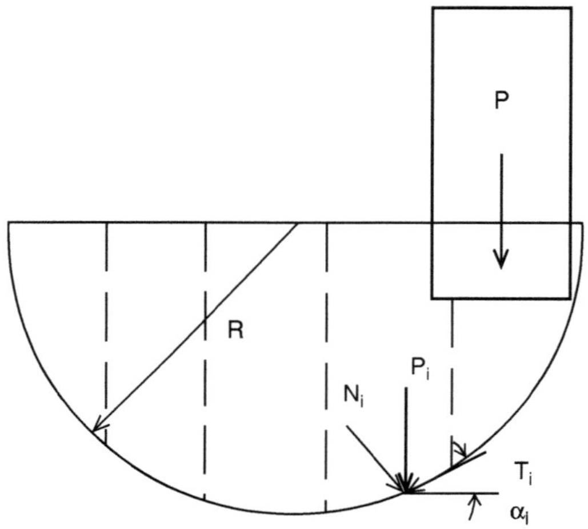

Fig. 4.29. Deep shear

In the first case the loss of stability occurs by a movement parallel to horizontal surface. An appreciation of strength is usually made by a calculation of a factor of stability as:

$$K_s = (fP + R_a)/Q \tag{4.73}$$

where Q is a shearing force, f – coefficient of friction, P – weight of the structure, R_a – resultant of active pressure computed by the relations (4.65), (4.67) and others of the previous paragraph or (4.72).

In the second case the loss of stability takes place by a movement along a cylindrical surface. The coefficient of stability can be calculated as a ratio of sums of moments of resistance and shearing forces:

$$K_s = \left(\sum_{i=1}^{n} (M_i)_{res} / \left(\sum_{i=1}^{n} M_i\right)_{sh}\right. . \tag{4.74}$$

To compute these sums we subdivide the soil massif by parts (blocks) and find for each of them normal and tangent forces as:

$$N_i = P_i \cos\alpha_i, \ T_i = P_i \sin\alpha_i. \tag{4.75}$$

With consideration of relations (4.75) expression (4.74) can be represented in the following way:

$$K_s = \left(\sum_{i=1}^{n} N_i \tan\varphi + cL\right) \Big/ \sum_{i=1}^{n} T_i \tag{4.76}$$

where c is a specific coherence, L – a length of slip arc.

4.2.9 Elementary Tasks of Slope Stability

Soil Has Only Internal Friction

We consider a slope inclined to the horizon under angle β (Fig. 4.30). Particle M on its surface has weight P. We decompose it in normal N and tangent T components. Force T' of friction resists to a movement of the particle. From the equilibrium condition we have:

$$P \sin\beta = \tan\varphi P \cos\beta$$

or

$$\tan\beta = \tan\varphi. \tag{4.77}$$

It means that ultimate angle β of a slope in quicksand is equal to its angle of internal friction φ.

Influence of Filtration Pressure

The angle of internal friction depends on hydrodynamic pressure D of a water in a condition of its filtration. In this situation shearing forces are (Fig. 4.31)

$$T = P \sin\beta, \ D = \gamma_w n i \sin\beta \tag{4.78}$$

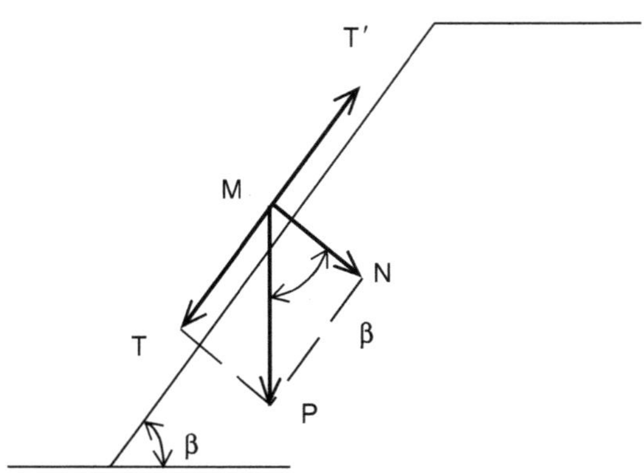

Fig. 4.30. Equilibrium of particle on slope surface

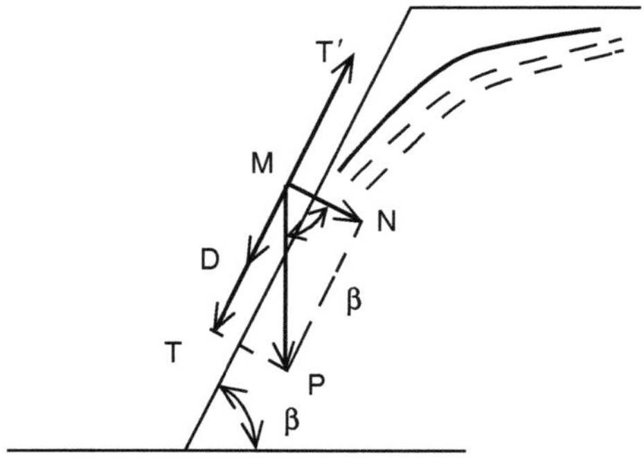

Fig. 4.31. Influence of filtration pressure

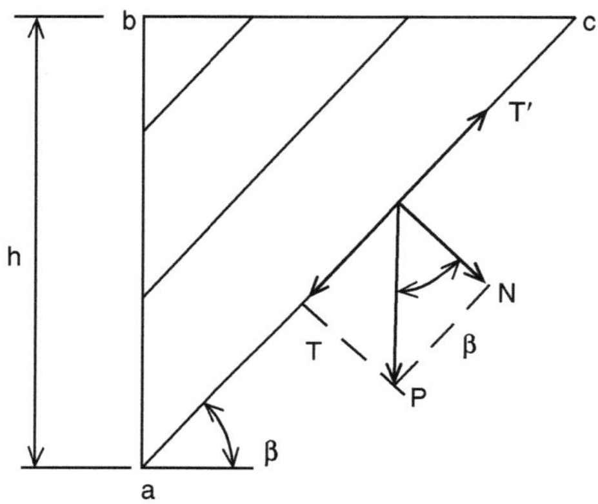

Fig. 4.32. Vertical slope of coherent soil

where γ_w is a specific weight of the water, n – porosity, $i \sin \beta$ – a hydraulic gradient. Resistance force is:

$$T' = P' \cos \beta \tan \varphi. \tag{4.79}$$

Here $P' = (\gamma_e)' i$ and $(\gamma_e)'$ is a specific weight of soil suspended in the water. With consideration of (4.78), (4.79) the stability factor is:

$$K_s = T'/(T + D) = (\gamma_e)' \tan \varphi / ((\gamma_e)' + \gamma_w n) \tan \beta. \tag{4.80}$$

Coherent Soil

Now we consider a vertical slope of coherent earth when slip surface is a plane (Fig. 4.32).

The acting force is self-weight P of sliding prism abc as:

$$P = 0.5 \gamma_e h^2 \cot \beta \tag{4.81}$$

from which

$$T = 0.5\gamma_e h^2 \cos\beta \tag{4.82}$$

The force of resistance is

$$T' = hc/\sin\beta.$$

In ultimate state $T = T'$ and with consideration of (4.82) we derive

$$0.5h^2\gamma_e \cos\beta = hc/\sin\beta \tag{4.83}$$

from which

$$c = (\gamma_e h/4)\sin 2\beta. \tag{4.84}$$

According to condition of ultimate equilibrium $\beta = \pi/4 - \varphi/2$ and at $\varphi = 0$ the slip plane makes with the horizon angle $\pi/4$. Taking this into account we find ultimate height of the vertical slope as:

$$h = 4c/\gamma_e.$$

4.2.10 Some Methods of Appreciation of Slopes Stability

Rigorous Solutions of Ultimate Equilibrium Theory

A rigorous solution of slope stability takes into account both the angle of internal friction and the coherence. Two main cases should be considered.

1. Maximum vertical pressure is given by relation (4.61) which corresponds to plane slope. Here some special tables are also used at different β, φ and dimensionless ultimate pressure σ_o can be found. Then the whole ultimate pressure with consideration of coherence can be written as follows:

$$p_u = \sigma_o + c\cot\varphi,$$

2. A slope in its ultimate state can support on its horizontal surface uniformly distributed load with intensity

$$p_* = 2c\cos\varphi/(1 - \sin\varphi).$$

This value can be considered as an action of an equivalent soil's layer with a height

$$h = 2c\cos\varphi/\gamma_e(1 - \sin\varphi).$$

When c and φ both are not equal to zero a construction of most equally stable slope may be fulfilled by the Sokolovski's method of dimensionless coordinates

$$x = Xc/\gamma_s, \;\; y = Yc/\gamma_s \tag{4.85}$$

beginning from the top of the slope.

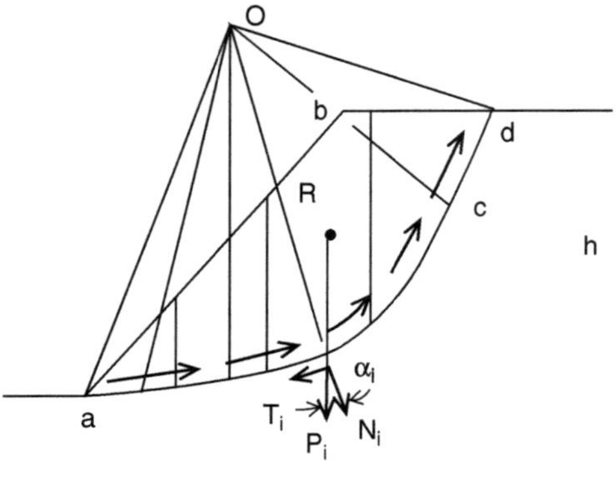

Fig. 4.33. Circular cylindrical surface

Method of Circular Cylindrical Surfaces

The method consists in a determination of a stability coefficient of natural slope for the most dangerous slip surfaces. In practice they are taken circular cylindrical and by a selection of the centre of the most dangerous one (for which K_s has minimum) is found.

Let the centre be in a point O (Fig. 4.33). We draw from it through the lower point an arc of slip and construct the equilibrium equation for massif abd. For this purpose we divide it by vertical cross-sections in n parts and use condition $\Sigma M = 0$ as

$$\sum_{i=1}^{n} T_i R - \sum_{i=1}^{n} N_i R \tan\varphi - cLR = 0. \tag{4.86}$$

Excluding from (4.86) R we have:

$$K_s = \left(\sum_{i=1}^{n} N_i \tan\varphi + cL\right) \Big/ \sum_{i=1}^{n} T_i. \tag{4.87}$$

To receive the most dangerous surface we behave in the following way (Fig. 4.34). We begin with case $\varphi = 0$ and find point O using angles β_1, β_2 from Table D1 of Appendix D. Then we put points $O_1, O_2, \ldots$ at equal distances and compute for each of them c-values according to (4.86) for consequent sliding surface. c_{max} corresponds to the most dangerous slope.

A simplification of this method was made by Prof. M. Goldstein according to whom

$$K_s = A\tan\varphi + Bc/\gamma_e h \tag{4.88}$$

where coefficients A, B must be taken from Table D2 of Appendix D. It is not difficult to find h (Fig. 4.35) as

$$h = cB/\gamma_e(K_s - A\tan\varphi) \tag{4.89}$$

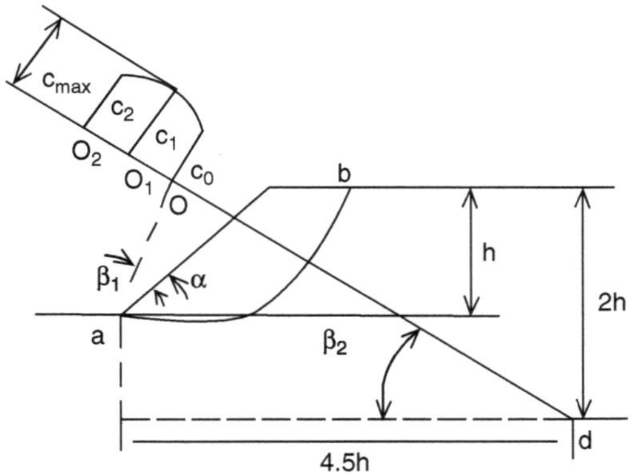

Fig. 4.34. Search for the most dangerous sliding surface

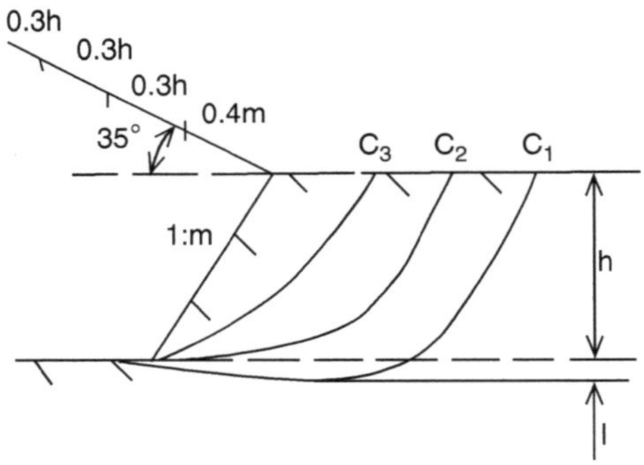

Fig. 4.35. Method of M. Goldstein

Method of Equally Stable Slopes by Approach of Professor Maslov

The method is based on the supposition that at the same pressure angle φ of resistance to shear in laboratory tests is linked with angle of repose ψ in natural conditions as

$$\tan\psi = \tan\varphi + c/\gamma_e H. \tag{4.90}$$

To construct a profile of a stable slope we divide it on a row of layers (Fig. 4.36) and compute for each of them the pressure of the soil on a lower plane and angle of shear by (4.90) with consideration of stability coefficient as follows:

$$\tan\psi = (\tan\varphi + c/p)/K_s. \tag{4.91}$$

The profile of such a slope with computed values of ψ beginning from the lower layer is given in Fig. 4.36.

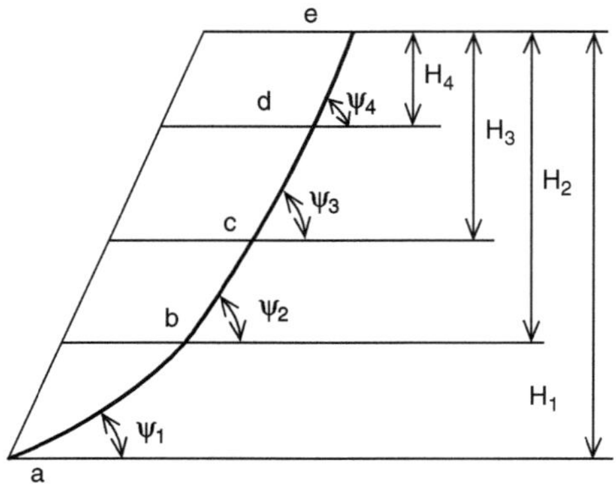

Fig. 4.36. Profile of equally stable slope

Method of Leaned Slopes

This method is used for an appreciation of landslide stability at fixed slip slopes and stability coefficient is computed according to (4.87). For a choice of a place of prop structure a pressure of a landslide must be found by this way. The massif is divided by parts (blocks) and for each of them the slip surface is a plane. According to the equilibrium condition for each of them (Fig. 4.33)

$$R_i + N_i \tan\varphi - cL_i - T_i = 0 \tag{4.92}$$

we have

$$R_1 = P_1 \sin\alpha_1 - P_1 \cos\alpha_1 \tan\varphi_1 - c_1L_1,$$
$$R_2 = P_2 \sin\alpha_2 - P_2 \cos\alpha_2 \tan\varphi_2 - c_2L_2 + R_1 \cos(\alpha_1 - \alpha_2)$$

or

$$R_i = P_i \sin\alpha_i - P_i \cos\alpha_i \tan\varphi_i - c_iL_i + R_{i-1} \cos(\alpha_{i-1} - \alpha_i) \tag{4.93}$$

where R_{i-1} is the projection of landslide pressure of preceding part on the direction of slip of the block in the consideration in the point with R_{min} corresponds to the place of the structure.

4.3 Axisymmetric Problem

4.3.1 Elastic-plastic and Ultimate States of Thick-walled Elements

Sphere

We begin with a sphere (Fig. 3.23) and computing difference of stresses $\sigma_\theta - \sigma_\rho$ from (3.114) and equalling it to σ_{yi} we find the difference of pressures which corresponds to the beginning of plastic deformation at $\rho = a$ ($q > p$ is everywhere in this paragraph) at $\beta = b/a$ as follows

$$(q - p)_{yi} = 2\sigma_{yi}(1 - \beta^{-3})/3. \quad (4.94)$$

When $q - \rho > (q - p)_{yi}$ we have two zones – an elastic in $c \le \rho \le b$ where

$$\sigma_\rho = C_1 + C_2/\rho^3, \sigma_\theta = C_1 - C_2/2\rho^3$$

and a plastic one at $a \le \rho \le c$. In the latter we determine from (2.80), boundary condition $\sigma_\rho(a) = -q$ and yielding demand $\sigma_\theta - \sigma_\rho = \sigma_{yi}$ – expressions

$$\sigma_\rho = -q + 2\sigma_{yi}\ln(\rho/a), \sigma_\theta = -q + \sigma_{yi}(1 + 2\ln(\rho/a)). \quad (4.95)$$

Constants C_1, C_2 can be excluded according to conditions $\sigma_\rho(b) = -p$ and $\sigma_\theta - \sigma_\rho = \sigma_{yi}$ at $\rho = c$. As a result we have in the elastic zone

$$\sigma_\rho = -p + 2c^3\sigma_{yi}(1 - b^3/\rho^3)/3b^3, \sigma_\theta = -p + 2c^3\sigma_{yi}(1 + b^3/2\rho^3)/3b^3. \quad (4.96)$$

From the compatibility law for stresses at $\rho = c$ we find the dependence of $p - q$ on c and its ultimate value at $c = b$ as follows:

$$q - p = 2\sigma_{yi}(1 - c^3/b^3 + 3\ln(c/a)), \ (q - p)_u = \sigma_{yi}\ln\beta. \quad (4.97)$$

Cylinder

It can be considered in the same manner. From (3.114) we find the difference of pressures at which the first plastic strains appear as

$$(q - p)_{yi} = \sigma_{yi}(1 - \beta^{-2})/2. \quad (4.98)$$

At $(q - p)_{yi} \le (q - p)$ we have in the elastic zone $c \le r \le b$

$$\sigma_r = -p + c^2\sigma_{yi}(1 - b^2/r^2)/2b^2, \sigma_\theta = -p + c^2\sigma_{yi}(1 + b^2/r^2)/2b^2.$$

The dependence of $q - p$ on c and its ultimate value at $c = b$ are

$$q - p = 0.5\sigma_{yi}(1 - c^2/b^2 + 2\ln(c/a)), \ (q - p)_u = \sigma_{yi}\ln\beta. \quad (4.99)$$

In plastic zone the expressions for stresses are similar to (4.95) and they can be written with a help of (2.67) as follows:

$$\sigma_r = -q + \sigma_{yi}\ln(r/a), \sigma_\theta = -q + \sigma_{yi}(1 + \ln(r/a)).$$

Comparing (4.94), (4.97)–(4.99) respectively we can conclude that a sphere demands the somewhat bigger difference of pressures for the beginning of yielding and twice of it at ultimate state than a cylinder.

Cone

Similarly the case of the cylinder (Fig. 3.24) the first plastic strains appear according to (3.118) at

$$(q - p)_{yi} = 0.5A\sigma_{yi} \sin^2 \psi / \cos \psi$$

where A is given by (3.120). At $q-p > (q-p)_{yi}$ we have in elastic and plastic zones respectively

$$\begin{aligned}
\frac{\sigma_\theta}{\sigma_\chi} &= -p + 0.5\sigma_{yi}\sin^2\upsilon \left(\cos\lambda/\sin^2\lambda \pm \cos\chi/\sin^2\chi \right. \\
&\quad \left. + \ln\left(\tan(\lambda/2)/\tan(\chi/2)\right)\right)/\cos\upsilon, \\
\sigma_\chi &= -q + \sigma_{yi}\ln(\sin\chi/\sin\psi), \sigma_\theta = -q + \sigma_{yi}(1 + \ln(\sin\chi/\sin\psi)).
\end{aligned}$$

The dependence of $q - p$ on angle υ at the border between elastic and plastic zones and the ultimate state are described by relations

$$\begin{aligned}
q - p &= 0.5\,\sigma_{yi}(1 + 2\ln(\sin\upsilon/\sin\psi) - \sin^2\upsilon(\cos\upsilon/\sin^2\upsilon) \\
&\quad + \ln(\tan(\lambda/2)/\tan(\upsilon/2))), (q - p)_u = \sigma_{yi}\ln(\sin\lambda/\sin\psi). \quad (4.100)
\end{aligned}$$

Let us now consider the yielding of the cone with initial angles ψ_o, λ_o when it is in the ultimate state. From (2.79) we derive the constant volume equation for displacement $V = u_\chi/\rho$ in following form:

$$du_\chi/u_\chi d\chi = -\cot\chi$$

with obvious solution

$$u_\chi = u_\chi(\psi)\sin\psi/\sin\chi.$$

But according to definition

$$u_\chi(\chi)/u_\chi(\psi) = d\chi/d\psi$$

that gives the integral which can be also found from the condition of constant volume of differences of spherical sectors

$$\cos\chi - \cos\chi_o = \cos\psi - \cos\psi_o \quad (4.101)$$

and instead of (4.100) we can write

$$(q - p)_u = 0.5\,\sigma_{yi}\ln(1 - (\cos\psi - \cos\psi_o + \cos\lambda_o)^2/\sin^2\psi).$$

Sokolovski /18/ investigated also the case of $\pi/2 \leq \lambda$ when two plastic zones (AOB and COD in Fig. 4.37) appear. This problem can be considered similarly to the previous one.

Particularly the ultimate state takes place at $\upsilon = \pi/2$ and for it we have also (4.100).

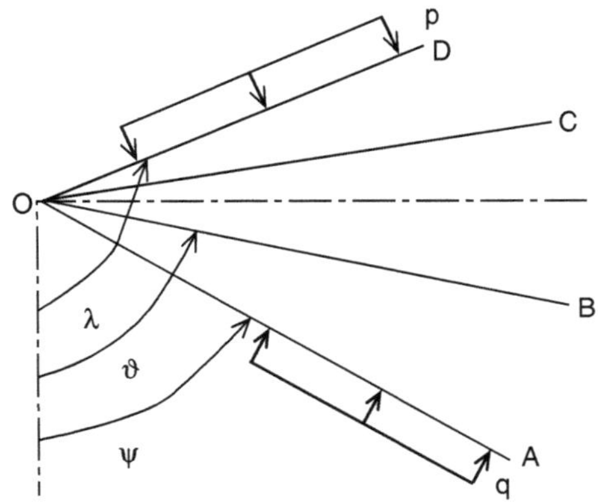

Fig. 4.37. Cone with big angle at apex

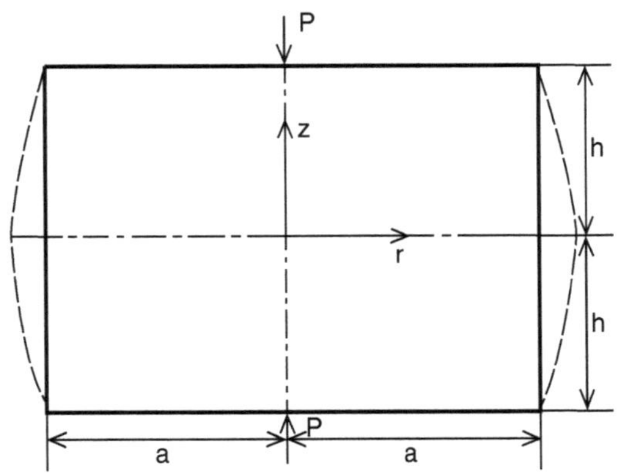

Fig. 4.38. Compression of cylinder

4.3.2 Compression of Cylinder by Rough Plates

L. Kachanov found upper load compressing a cylinder of height 2 h and radius a (Fig. 4.38). He proposed displacements in a form

$$u_r = U = Cr(1 - \beta z/h), u_z = V = V_o z/h$$

where β is a barrel factor, V_o – velocity of the plates and constant C can be found from the constant volume demand (see expressions (2.76)) as follows:

$$dU/dr + U/r + dV/dz = 0.$$

Then L. Kachanov uses the equality of power of external and internal forces as

$$P_* V_o = 2\pi a^2 \tau_{yi} \left(2a \int_0^{\eta} \int_0^{1} \varepsilon_e \rho d\rho d\xi + \int_0^{1} U(\rho)\, \rho d\rho \right). \quad (4.102)$$

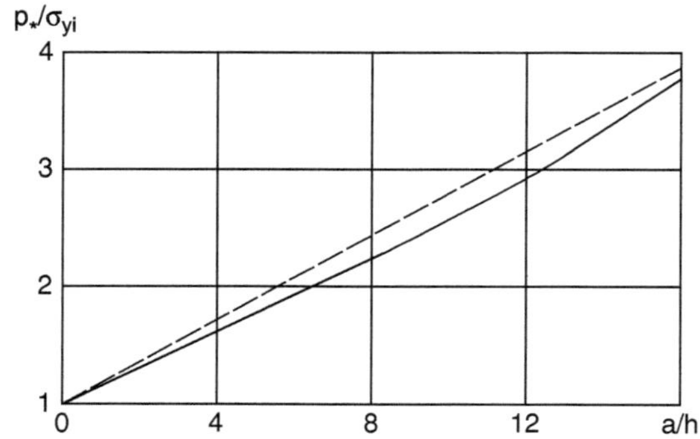

Fig. 4.39. Dependence of ultimate pressure on a/h

Here $\eta = h/a, \rho = r/a, \xi = z/a$ and equivalent strain ε_e is given by relation (2.25). Computing ε_r, ε_z, ε_θ, γ_{rz} according to (2.25), putting it into (4.102) we get on a complex expression. Parameter β should be found from the condition of minimum p_* where $p = P/\pi a^2$. The dependence of p_* on a/h is shown by solid curve in Fig. 4.39. The broken line in the figure corresponds to elementary solution $\beta = 0$ when $\varepsilon_e = \varepsilon_z = V_o/h, U = V_o a\rho/2h$ in form

$$p_*/\sigma_{yi} = 1 + 1/6\eta.$$

It is easy to see that the simple solution gives results near to the rigorous ones. In a similar manner the problems of stress state finding can be fulfilled for a neck in a bar at tension.

4.3.3 Flow of Material within Cone

Common Case

Similarly Sect. 4.2.2 we consider a flow within a cone (Fig. 3.6 where coordinates r, θ must be replaced by ρ, χ respectively). Above that we suppose here $\tau \equiv \tau_{\chi\theta}$ and that strains $\varepsilon_\chi = \varepsilon_\theta = -\varepsilon_\rho/2$ depend only on χ. Removing from (2.78) difference $\sigma_\rho - \sigma_\chi$ according to (2.65) and condition $\tau_e = \tau_{yi}$ we get on the first integral as

$$d\tau/d\chi = 2n\tau_{yi} - \tau \cot\chi - 4\sqrt{(\tau_{yi})^2 - \tau^2}$$

where n is a constant. Putting here representation (4.13) at $r = \rho, \theta = \chi$ we get equation

$$d\psi/d\chi = n/\cos 2\psi - 2 - 0.5 \cot\chi \tan 2\psi \qquad (4.103)$$

that can not be solved rigorously. Sokolovski /18/ gave diagrams $\lambda(n)$ and $\psi(\chi)$ for $\lambda \leq 40°$.

These curves for $0 \leq \lambda \leq \pi$ are represented by solid lines in Figs. 4.40 and 4.41.

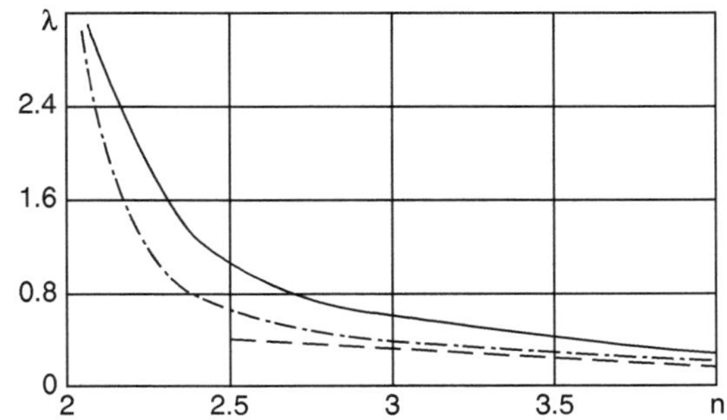

Fig. 4.40. Diagram $\lambda(n)$

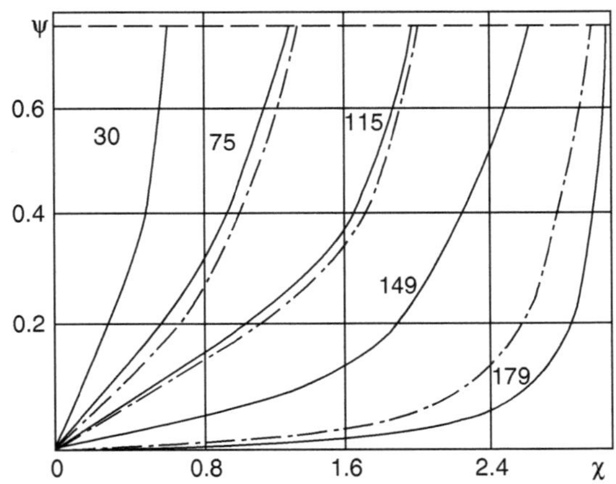

Fig. 4.41. Diagrams $\psi(\chi)$ at different λ

Now from the second equation (2.77) and representation (4.13) at $\tau_e = \tau_{yi}$ we derive

$$\sigma_\chi = F(\rho) - 3\tau_{yi} \int \sin 2\psi d\chi$$

where F is a function of ρ. Computing from (4.13) stress σ_ρ and putting it into the first (2.77) we find with consideration of (4.103) $F(\rho)$ and hence stresses depending on constant C.

$$\sigma_\chi = C - \tau_{yi}\left(2n \ln \rho + 3 \int \sin 2\psi d\chi\right), \ \sigma_\rho = \sigma_\chi + 2\tau_{yi} \cos 2\psi. \qquad (4.104)$$

Case of Big n

In /18/ Sokolovski proposed some simplifications as in Sect. 4.2.2 and got results for small λ

$$\lambda = 1/2n,\ \tau = \tau_{yi}\chi/\lambda,\ \sigma_\rho - \sigma_\chi = (2\tau_{yi}/\lambda)\ln(a/\rho),$$
$$\sigma_\rho = 2\tau_{yi}\left(\lambda^{-1}\ln(a/\rho) + 2\sqrt{1-(\chi/\lambda)^2}\right).$$

Diagram $\lambda(n)$ according to the first of these expressions is drawn in Fig. 4.40 by broken line and we can see that it is valid only for very small λ.

Approximate Approach

If we neglect in (4.103) the last member we can integrate the equation rigorously as follows:

$$\chi = 0.5\left(n(n^2-4)^{-0.5}\tan^{-1}\left(\sqrt{(n+2)/(n-2)}\tan\psi\right) - \psi\right).$$

(interrupted by points lines in Fig. 4.41). At $\psi = \pi/4$ this relation gives expression for $\lambda(n)$ – broken-solid curve in Fig. 4.40. The vicinity of the curves at the same λ in Fig. 4.41 allows to use this approach for further study From (4.18), (4.104) and integral static equation

$$\int_0^\lambda \sigma_\rho(a,\chi)\sin 2\chi d\chi = 0 \qquad (4.105)$$

we have approximately (see also Appendix E)

$$\begin{matrix}\sigma_\chi \\ \sigma_\rho\end{matrix} = \tau_{yi}\left(2n\ln(a/\rho) - J/8\sin^2\lambda - 0.375n\ln(n-2\cos 2\psi)\right.$$
$$\left.\begin{matrix}-0.75 \\ +1.25\end{matrix} \times \cos 2\psi\right)$$

where (solid line in Fig. 4.42)

$$J = \int_0^\lambda (10\cos 2\psi - 3n\ln(n-2\cos 2\psi))\sin 2\chi d\chi.$$

If we suppose $\sigma_\rho(a,\lambda) = \sigma_\chi(a,\lambda) = -q_*$ then we find

$$q_* = \tau_{yi}(3n\ln n + J/\sin^2\lambda)/8 \qquad (4.106)$$

and this expression will be used later in Chap. 5.

4.3.4 Penetration of Rigid Cone and Load-bearing Capacity of Circular Pile

These problems can be solved in the same manner as in Sect. 4.2.4. The basic relations of the previous subparagraph are valid here but instead of

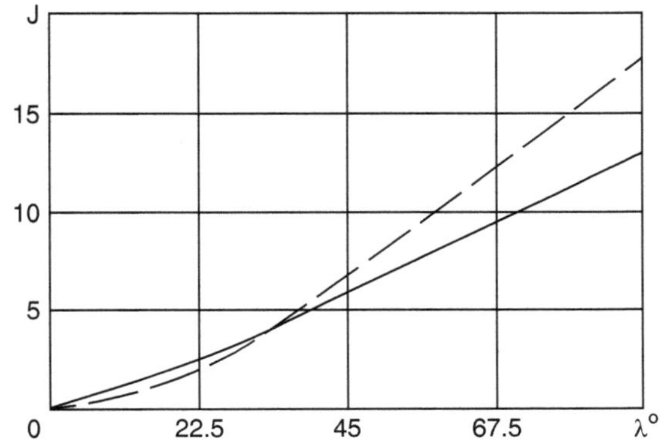

Fig. 4.42. Diagram J(λ)

(4.105) we must use similar to (4.20) (Fig. 4.9 with consequent replacement of coordinates) integral static laws for axisymmetric problem

$$\begin{aligned} p_* b^2 &= -2a^2 \int_0^{\lambda} (\sigma_\rho(a,\chi)\cos\chi + \tau(a,\chi)\sin\chi)\sin\chi d\chi, \\ P_*/\pi &= p_* b^2 + 2\sin\lambda \int_a^{a+1} (\sigma_\chi(\rho,\lambda)\sin\lambda + \tau(\rho,\lambda)\cos\lambda)\rho d\rho. \end{aligned} \tag{4.107}$$

As a result we find

$$\begin{aligned} P_*/\pi \ &= p_* b^2 + \tau_{yi}(l(l+2a)(J_1/4 + \cos\lambda\sin\lambda + n(1-(3/8)\ln n)\sin^2\lambda) \\ &-2n(l+a)^2\ln(l/a+1)\sin^2\lambda). \end{aligned} \tag{4.108}$$

Here

$$J_1 = \int_0^{\lambda} (10\cos 2\psi - 3n\ln(n - 2\cos 2\psi)\cos\chi + 8\sin 2\psi\sin\chi)\sin\chi d\chi.$$

For λ near to π and $a \to \infty$ (the option of a circular pile) we compute that it is valid

$$P_*/\pi = p_* b^2 + 2\tau_{yi} bl(1 + J_1/4\sin\lambda). \tag{4.109}$$

The calculations at $n = 2.045$, $\lambda = 179°$ give $J_1 \approx 6.5$ and hence ratio $J_1/4\sin\lambda$ is big (due to an approximative character of the above theory). Neglecting this member we have a simple result in a safety side as following

$$P_* = \pi b(p_* b + 2\tau_{yi} l). \tag{4.110}$$

4.4 Intermediary Conclusion

An importance of the results according to the scheme of the perfect plastic body is difficult to overestimate. They give ultimate loads and therefore – the moment of structures destruction. We can also notice that a procedure of computations is much simpler that by the Theory of Elasticity methods. However the solutions of this chapter can be used first of all for the materials with big yielding part of stress-strain diagram or small hardening (for soils – with distinctive angle of internal friction and cohesiveness). Above that the process of deformation and fracture between elastic and plastic stages is unknown. This gap can be removed in computations according to equations of the hardening body which is considered in the following two chapters where unsteady non-linear creep and damage are also taken into account.

5

Ultimate State of Structures at Small Non-Linear Strains

5.1 Fracture Near Edges of Cracks and Punch at Anti-Plane Deformation

5.1.1 General Considerations

Equilibrium equation (2.56) is satisfied if we take

$$\tau_r = -\partial W/r\partial\theta, \tau_\theta = \partial W/\partial r$$

where W is a potential function that can be written in form $W = Kr^s f(\theta)$ which gives relations for stresses

$$\begin{gathered}\tau_r = -Kr^{s-1}f'(\theta),\ \tau_\theta = Ksr^{s-1}f(\theta),\\ \tau_e = Kr^{s-1}\sqrt{(sf)^2 + (f')^2}.\end{gathered} \tag{5.1}$$

Since /21/ $\tau_i\gamma_i$ is proportional to r^{-1} we find with a help of (2.53) at $\alpha = 0$ $s = m/(m+1)$ and according to (2.57), (5.1) – derive following values

$$\begin{gathered}u_z = -\Omega(t)(m+1)K^m r^{1/(m+1)}((sf)^2 + (f')^2)^{(m-1)/2}f',\\ \tau_\theta = Kmr^{-1/(m+1)}f/(m+1).\end{gathered} \tag{5.2}$$

From compatibility expression for strains (2.58) we have differential equation for $f(\theta)$ as /16/

$$f'' + (1 + m(m-1)(((m+1)f')^2 + f^2)/(((m+1)f')^2 + mf^2))F/(m+1)^2 = 0. \tag{5.3}$$

At $m = 1$ we have from (5.3) results (3.10) and (3.19).

In order to get a rigorous solution we put in (5.3)

$$f = e^z, f' = e^z U, f'' = e^z(U^2 + U')$$

that gives (C is a constant)

$$U = \left(\pm\sqrt{(1+m)^2 + 4m\tan^2(C-\theta)} - (1+m)\right)/2(1+m)\tan(C-\theta). \quad (5.4)$$

On the other hand we have $\ln(f/D) = Ud\theta$ (D is also a constant) and with consideration of (5.4) as well as supposing $\tan(C-\theta) = \sqrt{\eta}$ we receive

$$4(m+1)d\ln(f/D) = \left(1 + m - \left(\pm\sqrt{(m+1)^2 + 4m\eta}\right)\right) d\eta/\eta(1+\eta) \quad (5.5)$$

and after integration we find (see Appendix F)

$$f = D(\surd - m + 1)/(\surd + m - 1)^{(m-1)/4(m+1)}$$
$$\times ((\surd + m + 1)/(\surd - m - 1))^{1/4}\sqrt{\sin(C-\theta)} \quad (5.6)$$

where

$$\surd = \sqrt{(1+m)^2 + 4m\tan^2(C-\theta)}.$$

5.1.2 Case of Crack Propagation

For this task from boundary conditions $f(0) = 1, f(\pi) = 0$ (Fig. 3.4 and relations (5.1)) we compute $C = 0$ and the value of another constant

$$D_1 = m^{m-1/4/(1+m)}/(1+m)^{0.5}. \quad (5.7)$$

From Fig. 5.1 where according to expressions (5.1) and (5.6) diagrams $\tau_r(\theta)$, $\tau_\theta(\theta)$ are constructed for $m = 1, 3, 7$ by solid, broken and interrupted by points lines we can see that with an increase of m the role of one component of τ_e in the certain part of the plane is growing.

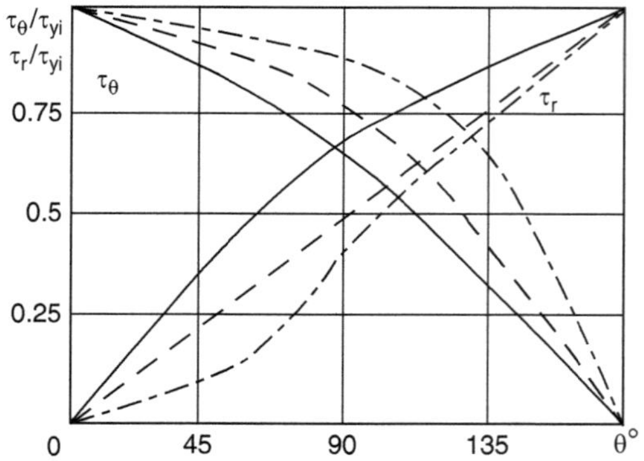

Fig. 5.1. Diagram of shearing stresses near crack end

In order to find the value of K we compute integral $J = d\Pi/dl$ where

$$d\Pi = \int_0^{dl} \tau_\theta(r,0) u_z(dl - r, \pi) dr$$

is a free energy of crack propagation. Taking into an account u_z and τ_θ from (5.2) we calculate

$$J = \Omega(t) K^{m+1} m (f'(\pi))^m \int_0^1 (\xi/(1-\xi))^{1/(1+m)} d\xi.$$

According to /21/ a value of J in the similar problem for tension does not depend on the properties of a material. Assuming this supposition also for our task we can equal J to its meaning at m = 1 that gives

$$K^{m+1} = \tau_o^2 \pi l / 2G\Omega(t)/f'(\pi)/^m I(m)$$

where

$$I(m) = (m+1)\Gamma((2+m)/(1+m))\Gamma((2m+1)/(m+1)).$$

Here $\Gamma()$ is gamma-function and $f'(\pi)$ must be taken from solution (5.6). Computing τ_e by the third expression (5.1) we receive the following equation for $r(\theta)$ at τ_e = constant

$$2G\Omega(\tau_e)^{1+m} r/(\tau_o)^2 l = \pi \left(\sqrt{(sf)^2 + f'^2}\right)^{m+1} /I(m)/f'(\pi)/^m. \qquad (5.8)$$

From Fig. 5.2b where condition (5.8) is represented by solid, broken and interrupted by points lines for m = 1, 3, 7 respectively for the elastic-plastic

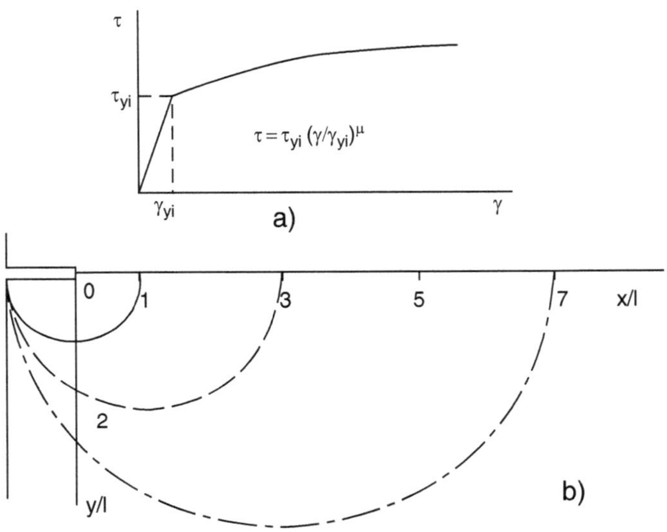

Fig. 5.2. Plastic zones for non-linear material

material with diagram in Fig. 5.2a (see Appendix F) we can see that with the growth of m plastic zone increases and moves out of the crack. It confirms the solution of Sect. 4.1.2 for a perfect plastic body.

5.1.3 Plastic Zones Near Punch Edges

In this case we have boundary conditions $f(0) = 0, f(\pi) = 1$ and from (5.6) we find $C = 0$ and D_2 in Appendix F. It is shown there that in the previous solution angle θ must be replaced by value $\pi - \theta$ and diagram in Fig. 5.1 should be reflected relatively to axis $\theta = \pi/2$.

5.2 Plane Deformation

5.2.1 Generalization of Flamant's Problem

Common Solution

As in Sect. 3.2.3 we suppose $\sigma_\theta = \tau_{r\theta} = 0$ (Fig. 3.11) that allows to find from the first static equation (2.67)

$$\sigma_r = f(\theta)/r \tag{5.9}$$

and according to rheological law (2.66) at $\sigma_e = \sigma_x = \sigma_r$ we have

$$\varepsilon_r = -\varepsilon_\theta = 0.75\Omega(t)r^{-m}g(\theta) \tag{5.10}$$

wherein $g(\theta) = f^m$. From (2.66) at $\alpha = 0$ and condition $\gamma_{r\theta} = 0$ compatibility expression (2.71) becomes

$$g'' + \beta^2 g = 0. \tag{5.11}$$

Here $\beta = \sqrt{m(2 - m)}$. The solution of (5.11) depends on the value of m /18/. If $m = 2, m > 2$ and $m < 2$ then we have respectively

$$g = C\theta + D,$$

$$g = C\cos h\beta\theta + D\sinh\beta\theta, g = C\cos\beta\theta + D\sin\beta\theta.$$

The constants C D can be found from integral static laws (see Fig. 3.11) as

$$P\cos\alpha_o = -\int_{-\lambda}^{\lambda} r\sigma_r\cos\theta d\theta,\ P\sin\alpha_o = -\int_{-\lambda}^{\lambda} r\sigma_r\sin\theta d\theta. \tag{5.12}$$

Particular Cases

At $\lambda = \pi/2$ and $\alpha_o = 0$ (see Fig. 3.11) we compute for $m = 2$ $C = 0$. $D = (P/2)^2$ and stress

$$\sigma_r = P/2r$$

does not depend on angle θ.

In a similar way cases $m > 2, m < 2$ can be studied. Taking into an account the symmetry condition we receive respectively

$$\sigma_r = D_1(\cos h\beta\theta)^{\mu}/r, \sigma_r = D_2(\cos\beta\theta)^{\mu}/r. \tag{5.13}$$

Constants D_1, D_2 can be found from expressions (5.12). At $m = 1$ and $\alpha_o = 0$ we receive the Flamant's result (3.17).

For practical purposes it is interesting to establish the dependence of stress σ_y on angle θ. From (2.72), (5.12), (5.13) we have for $m = 1, 2, 4$ respectively

$$\begin{gathered}\sigma_y(1) = -(2P/\pi y)\cos^4\theta,\ \sigma_y(2) = -(P/2y)\cos^3\theta \\ \sigma_y(4) = -0.4(P/y)\left(\cos h2\sqrt{2}\theta\right)^{1/4}\cos^3\theta.\end{gathered} \tag{5.14}$$

Corresponding diagrams $/\sigma_y(\theta)/$ are constructed in Fig. 5.3 by solid, broken and interrupted by points curves. We can see that with an increase of m the stress distribution is more even.

Comparison of Results

In order to appreciate the results we compare for the case $m = 1$ the distribution of stresses σ_y along vertical axis under the concentrated load as well as centres of the punch and uniformly placed load where we have according to the first relation (5.14) (solid line in Fig. 5.4), and expressions (3.95) (broken curve in the figure), (3.52) (interrupted by points line) respectively

$$\begin{gathered}\sigma_y/p = -4l/\pi y,\ \sigma_y/p = -2(1 + 2(y/l)^2)/\pi(1 + (y/l)^2)^{3/2}, \\ \sigma_y/p = -2(\tan^{-1}(l/y) + (y/l)/(1 + (y/l)^2))/\pi.\end{gathered} \tag{5.15}$$

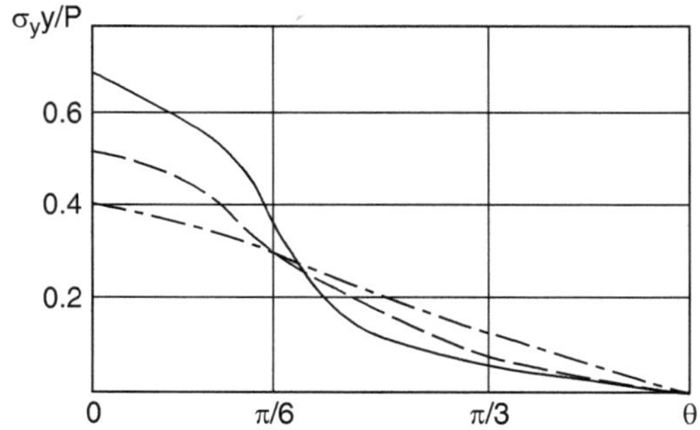

Fig. 5.3. Distribution of stresses at different m

Here $p = P/2l$ and we can see from Fig. 5.4 that at $y > 3l$ the curves practically coincide. Since with the growth of non-linearity the stress distribution becomes more uniform we can expect that solutions (5.13) can replace other forms of pressure on the foundation at least at $y > 3l$.

Case of Horizontal Force

The results of the previous subparagraph can be used here for the case $\lambda = \alpha_o = \pi/2$ if we compute angle θ from horizontal direction when we have for $m = 1, 2, 4$ respectively (solid, broken and interrupted by points lines in Fig. 5.5)

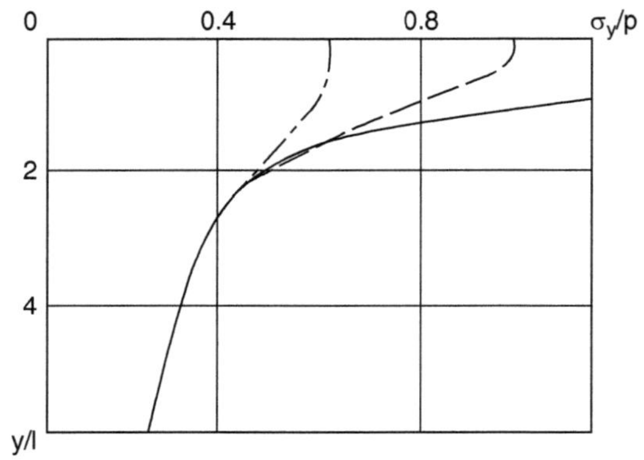

Fig. 5.4. Distribution of stresses under different loads

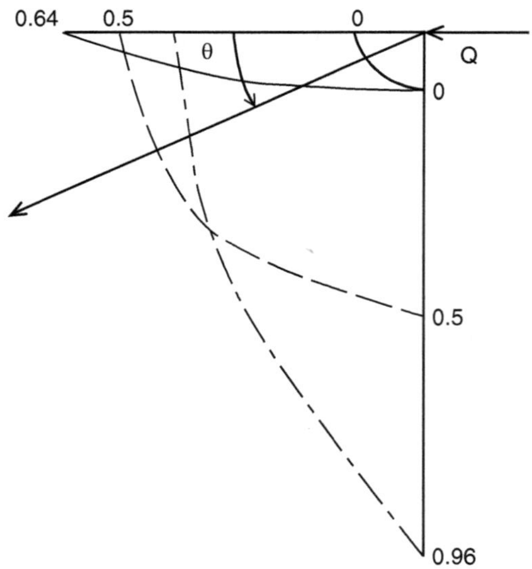

Fig. 5.5. Distribution of stresses due to horizontal force

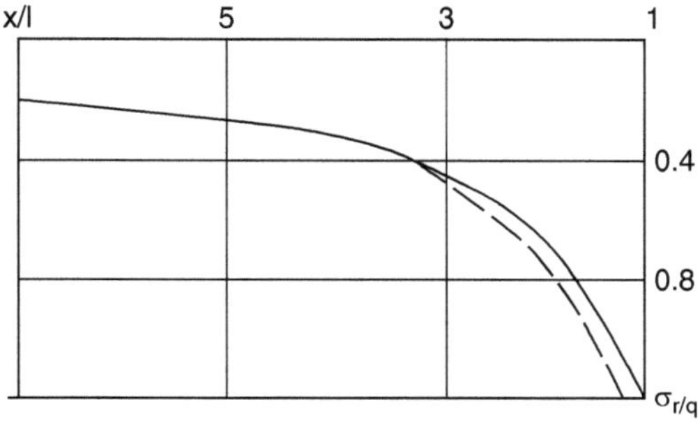

Fig. 5.6. Distribution of stresses at different loadings

$$\sigma_r = -2(Q/\pi r)\sin\theta,\ \sigma_r = -Q/2r,\ \sigma_r = -0.4Q\left(\cos h2\sqrt{2}\theta\right)^{1/4}/r. \quad (5.16)$$

In order to give to the results just received a practical meaning we compare for the case m = 1 the distribution of σ_r along axis x according to the first relation (5.16) and expressions (3.101), (3.109) as follows

$$\sigma_r = -4ql/\pi x, \sigma_r = -4q((x/l)^2 - 1)^{-1/2}/\pi.$$

where q = Q/2l, and from Fig. 5.6 in which the corresponding diagrams are given by solid and broken lines we can see that the curves are near to each other and practically coincide at x/l > 3. So, we can use results (5.16) in a non-linear state at least out of this district.

5.2.2 Slope Under One-Sided Load

General Relations

For the purpose of this paragraph we rewrite (2.66) at $\alpha = 0, \gamma_{r\theta} \equiv \gamma$ in following form (see also (2.30))

$$\sigma_r - \sigma_\theta = 4\omega(t)(\gamma_m)^{\mu-1}\varepsilon_r, \tau = \omega(t)(\gamma_m)^{\mu-1}\gamma \quad (5.17)$$

where γ_m is a maximum shearing stress

$$\gamma_m = \sqrt{(\varepsilon_r - \varepsilon_\theta)^2 + \gamma^2}$$

linked with τ_e by the law similar to (2.30) as

$$\tau_e = \omega(t)(\gamma_m)^\mu. \quad (5.18)$$

Putting (5.17) and similar to (4.13) representations for strains

$$\varepsilon_r = 0.5\gamma_m \cos 2\psi, \gamma = \gamma_m \sin 2\psi \quad (5.19)$$

into (2.68) and the third equation (3.21) we get on the system

$$d((\gamma_m)^\mu \sin 2\psi)/d\theta + 2(\gamma_m)^\mu \cos 2\psi = 0, \quad (5.20)$$

$$d(\gamma_m \cos 2\psi)/d\theta = 2\gamma_m \sin 2\psi + C_1 \quad (5.21)$$

where C_1 is a constant. At $\mu = 1$ we have from (5.20), (5.21) the solution of Sect. 3.2.1.

Fulfilling the operations in (5.20), (5.21) and excluding γ_m we receive the second order differential equation which is not detailed in /18/. Replacing in it

$$\Theta = -d\theta/d\psi \tag{5.22}$$

we find the first order differential equation

$$(\tan 2\psi)d\Theta/\Theta d\psi = 2(1-\Theta)((\Theta-1)/\mu + 1 - 2/\Psi) \tag{5.23}$$

in which

$$\Psi = 1 - (1-\mu)\sin^2 2\psi. \tag{5.24}$$

Sokolovski /18/ gave curves $\psi(\theta), \tau_e(\theta)$ and $\max\tau_e(\lambda)$ for $\mu = 1/3, \lambda < \pi/4$. The latter is shown by pointed line in Fig. 5.7 (we calculated it till $\lambda = \pi/3$). Solid curves 1 and 0 refer to expression (3.25) (for a linear material at $\theta = 0$ when $\lambda < \pi/4$ and $\lambda > \pi/4$) and 4.11. So we can conclude that the Sokolovski's solution is wrong at $\lambda < \pi/4$.

Here we integrate (5.23) by the finite differences method at boundary condition $\Theta(0) = 1$. Then we use (5.22) at border demand $\theta(0) = \lambda$. Another similar condition $\theta(\pi/4) = 0$ allows to choose ratio $(1-\Theta)/\tan 2\psi$ in point $\Theta(0) = 1$. The calculations were made by a computer.

Results of Computation

Firstly we consider case $\mu = 1$ when we have (solid curves in Fig. 5.8)

$$\tan 2\psi = (\cos 2\theta - \cos 2\lambda)/\sin 2\theta. \tag{5.25}$$

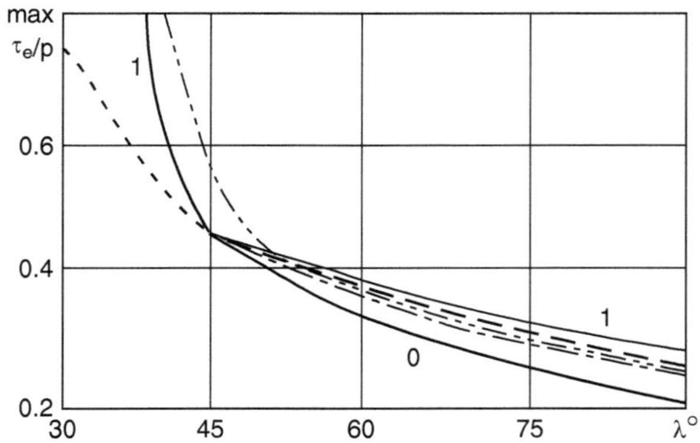

Fig. 5.7. Dependence of $\max\tau_e$ on λ

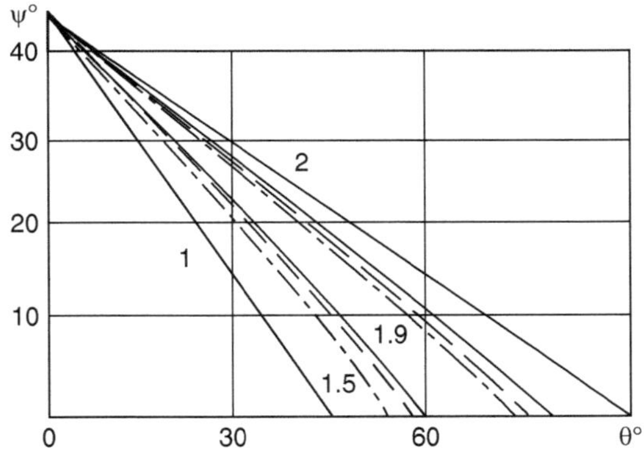

Fig. 5.8. Dependence of ψ on θ at different λ

At $\lambda = \pi/4$ and $\lambda = \pi/2$ we can receive from (5.25) straight lines 1,2

$$\psi = -\theta + \pi/4, \psi = -\theta/2 + \pi/4$$

respectively and the first is valid at any μ. Differentiating (5.25) we find

$$\Theta = (1 + \cos^2 2\lambda - 2\cos 2\lambda \cos 2\theta)/(1 - \cos 2\lambda \cos 2\theta). \tag{5.26}$$

or after transformations – as function of ψ with upper sign for $\lambda > \pi/4$ and lower one for $\lambda < \pi/4$:

$$\Theta = 1 \pm (\sin 2\psi)(\tan^2 2\lambda + \sin^2 2\psi)^{-1/2}.$$

The approximate calculations reveal good agreement with (5.25), (5.26). It allows to use the finite differences method for another μ. The curves for $\mu = 1/2$ and $\mu = 1/4$ for $\Theta_o = 1.5$. $\Theta_o = 1.9$ respectively are near to solid lines in Fig. 5.8.

When function $\psi(\theta)$ is known a value of τ_e can be found from equations following of (3.21), (4.13) and boundary conditions for σ_θ (see Sect. 3.2.1) as

$$d\tau_e/\tau_e d\theta + 2\,(d\psi/d\theta + 1)\cot 2\psi = 0,$$
$$p = 4\int_0^\lambda \tau_e \sin 2\psi d\theta.$$

Combining these expressions we find for $\max\tau_e = \tau_e(0)$ at $\lambda \geq \pi/4$

$$p = 4\max\tau_e \int_0^\lambda \sin 2\psi \exp\left(-2\int_0^\theta (1 + d\psi/d\theta)\cot 2\psi d\theta\right) d\theta. \tag{5.27}$$

Computations for $\mu = 2/3, 1/2$ and $1/3$ show that diagrams $\max\tau_e(\lambda)$ are near the solid line 1 in Fig. 5.7. It can be explained by the absence of μ in

(5.27) and the vicinity of curves in Fig. 5.8 at different μ. It allows to use the solid lines in the latter figure for practical purposes.

In order to find the ultimate state we rewrite (5.18) with consideration of (1.45), (2.30) on axis $\theta = 0$ where $\varepsilon_\theta = \varepsilon_r = 0, \gamma = 2\varepsilon_1 \equiv 2\varepsilon > 0$

$$\Omega(t)(2\max\tau_e)^m = \varepsilon e^{-\alpha\varepsilon}. \tag{5.28}$$

Using the criterion $d\varepsilon/dt \to \infty$ we receive the values at critical state as

$$\varepsilon_* = 1/\alpha, \Omega(t_*) = (2\max\tau_e e\alpha)^{-m}. \tag{5.29}$$

Simple Solution

In order to find engineering relations we rewrite (5.17) in form of (2.66) at $\alpha = 0$ as follows /25/

$$\varepsilon_r = \Omega(t)(\tau_e)^{m-1}(\sigma_r - \sigma_\theta)/4, \gamma = \Omega(t)(\tau_e)^{m-1}\tau \tag{5.30}$$

and put them into the third (3.21) that gives

$$(\tau_e)^{m-1}((m-1)(\tau_e)^{-2}\tau'^2 + 1)(\tau'' + 4\tau) = C_2. \tag{5.31}$$

where C_2 is a function of t. Here and further the dependence on time is hinted. According to the symmetry condition $\tau'(0) = 0$ and taking this assumption for the whole wedge we receive from (5.31)

$$(\tau_e)^{m-1}(\tau'' + 4\tau) = C_2$$

which gives at $m = 1$ the solution of Sect. 3.2.1.

To exclude C_2 from (5.31) we differentiate it as follows

$$\begin{aligned}(m-1)((m-3)(\tau''+4\tau)\tau'^2 + (\tau_e)^2(3\tau''+4\tau))\tau'(\tau''+4\tau)\\ + (\tau_e)^2((m-1)\tau'^2 + (\tau_e)^2)(\tau''' + 4\tau') = 0.\end{aligned} \tag{5.32}$$

Here we again suppose $\tau' = 0$ in the whole wedge that gives from (5.32) $\tau''' = 0$ and with consideration of (3.21) as well as the same boundary conditions as in Sect. 3.2.2 for σ_θ, τ at $\pm\lambda$ we compute

$$\begin{aligned}\tau = 3p(\lambda^2 - \theta^2)/8\lambda^3,\ \sigma_r - \sigma_\theta = 3p\theta/4\lambda^3,\\ \begin{matrix}\sigma_\theta\\ \sigma_r\end{matrix} = 3p(\theta^3/3 + \theta x^{-\lambda^2}_{1-\lambda^2})/8\lambda^3 - p/2,\ \tau_e = 3p\sqrt{\theta^2 - (\theta^2-\lambda^2)^2}/8\lambda^3.\end{aligned} \tag{5.33}$$

From Fig. 5.7 we can see that interrupted by two points curve corresponding to (5.33) at $\theta = 0(\lambda > 1)$ and $\theta = \lambda(\lambda < 1)$ may be taken as the first approach.

5.2.3 Wedge Pressed by Inclined Rigid Plates

Engineering Relations for Particular Case

We considered the problem of a pressed wedge in Sects. 3.2.2 and 4.2.2 for elastic and plastic media. Here we study the task for a hardening at creep material and begin with the case of parallel moving plates (Fig. 3.6) at negligible compulsory flow /23/. From (3.28) we have at $u_\theta = -V(\theta)$

$$u_r = V', \varepsilon_r = \varepsilon_\theta = 0, \gamma = (V'' + V)/r \qquad (5.34)$$

and using (5.17), (2.67) we find

$$\tau = \omega(t)r^{-\mu}f(\theta),\ \sigma_r = \sigma_\theta = F(r) - \omega(t)r^{-\mu}(2-\mu)\int f(\theta)d\theta \qquad (5.35)$$

where F is a function of r and

$$f(\theta) = (V'' + V)^\mu.$$

Putting (5.35) into (2.68) we get on equality

$$\omega(t)\left(\mu(2-\mu)\int f(\theta)d\theta + f'(\theta)\right) = -r^{1+\mu}dF/dr$$

which is true if its both parts are equal to the same function of t, say n(t). This gives two expressions

$$F = A - mr^{-\mu}n, f'' + \mu(2-\mu)f = 0.$$

Taking into account the symmetry condition we write the solution of the latter equation as following:

$$f(\theta) = C\sin\beta\theta.$$

Here $\beta = \sqrt{\mu(2-\mu)}, n = 0$ and constant C will be determined further. So, using condition $\sigma_r(a,\lambda) = \sigma_\theta(a,\lambda)$ and the second integral static laws (3.32) we derive for $\theta = \lambda$

$$\tau_e = /\tau/(q/mB_o\beta)(a/r)^\mu \tan\beta\lambda, \sigma_r = \sigma_\theta = -q(1-(a/r)^\mu)B_o \qquad (5.36)$$

where

$$B_o = 1 - a(1-(1+l/a)^{1-\mu})/l(1-\mu).$$

At $l >> a$ we have again $B_o = 1$ and for $\mu = \beta = 1$ – solution (3.40), (3.41). Diagrams $\sigma_\theta(r), \max\tau_e(\lambda)$ for $\mu = 0.5$ are given by pointed curves in Figs. 3.7–3.9. At $\lambda \to 0$ we get the solution near to that in /17/.

To find the critical state for $\varepsilon \equiv \varepsilon_1 = \gamma/2 > 0$ we apply criterion of infinite strain rate $d\varepsilon/dt \to \infty$ at dangerous points $a = r, \theta = \lambda$ and we compute with a consideration of (5.36), (5.28)

$$\varepsilon_* = 1/\alpha, \Omega(t_*) = ((2\alpha eq/m\beta)\tan\beta\lambda)^{-m}.$$

To receive displacement V we use expressions for f(θ) which give equation

$$V'' + V = C^m \sin^m \beta\theta$$

with

$$C = qa^{\mu}\Omega^{\mu}2^{1-\mu}/m\beta\cos\beta\lambda$$

which must be solved at boundary demands $V(0) = 0, V'(\lambda) = 0$ (stick condition) and $V(\lambda) = V_o$. The latter reveals dependence of V_o on load and time. For $m = 1, 2, 3, \ldots$ the equation may be easily solved e.g. At $m = 2$ we obtain

$$V_o = C^2\left(2 + \cos\sqrt{3}\lambda + \sqrt{3}\sin\sqrt{3}\lambda\tan\lambda - 3/\cos\lambda\right)/4.$$

Flow of Material Between Immovable Plates

Now we consider the case when only the compulsory flow takes place (a model of a volcano row) and here we have from (3.28)

$$u_r = U(\theta)/r, \varepsilon_\theta = -\varepsilon_r = U(\theta)/r^2, \gamma = dU/r^2d\theta. \tag{5.37}$$

According to (5.18) we find

$$\gamma_m = g(\theta)/r^2 \tag{5.38}$$

where

$$g = \sqrt{U'^2 + 4U^2}. \tag{5.39}$$

Using the representation similar to (5.19)

$$\varepsilon_r = (g/2r^2)\cos 2\psi, \gamma = (g/r^2)\sin 2\psi \tag{5.40}$$

we have from (5.37)

$$\ln(/U/ : D) = -2\int_0^\theta \tan 2\psi d\theta, \; g = -2U/\cos 2\psi. \tag{5.41}$$

Here D is a constant that will be found later. We must also notice that the solution satisfies stick condition $U(\lambda) = 0$.

Putting (5.40) into compatibility law following from (5.37) as

$$\partial\varepsilon_\theta/\partial\theta = \gamma$$

we receive equation

$$(g\cos 2\psi)' + 2g\sin 2\psi = 0 \tag{5.42}$$

which also gives the boundary condition

$$d\psi/d\theta = 1 \tag{5.43}$$

at $\theta = \lambda$. Above that we find from (5.42) expression for $g(\theta)$ as

$$\ln(g/D) = 2\int_0^\theta (d\psi/d\theta - 1)\tan 2\psi d\theta. \tag{5.44}$$

Now from (5.17) and (5.40) we derive expressions

$$\sigma_r - \sigma_\theta = 2\omega(t)r^{-2\mu}g^\mu \cos 2\psi, \tau = \omega(t)r^{-2\mu}g^\mu \sin 2\psi \tag{5.45}$$

which together with (2.68) give

$$(g^\mu \sin 2\psi)'' + 2(1-2\mu)(g^\mu \cos 2\psi)' + 4\mu(1-\mu)g^\mu \sin 2\psi = 0. \tag{5.46}$$

From (5.46), (5.42) we find after exclusion of $g(\theta)$ the second order differential equation for $\psi(\theta)$ which is not detailed in /18/. Replacing in it

$$\Phi = d\theta/d\psi \tag{5.47}$$

we derive the first order differential equation

$$(\cot 2\psi)d\Phi/d\psi = 2\Phi(\mu - 1 + 2\mu/\Psi - (1 + 2\mu^2/\Psi)\Phi + \mu^2\Phi^2/\Psi) \tag{5.48}$$

where

$$\Psi = \mu + (1-\mu)\cos^2 2\psi. \tag{5.49}$$

Expression (5.48) should be solved at different $\Phi(0) = \Phi_o$. Then we find function $\theta(\psi)$ at $\theta(0) = 0$ that corresponds to $\theta = \lambda, \Phi = 1$ at $\psi = \pi/4$ (Fig. 5.9 for $\mu = 0.5$) and finally we have $g(\theta)$ and $U(\theta)$.

Here we must notice that (5.41), (5.44) give different values of $g(\theta)$ since we use conditions $g(0)/D = 1, U(0)/D = 1$. To get the correct answer we recommend the following procedure. We compute $g_1/D, U_1/D$ according to expressions (5.41), (5.44) respectively. Then we find $/U_2//D$ by the first relation (5.41), calculate difference $U/D = (U_1 - /U_2//D)$ according to the second law (5.41). Since g and ψ in (5.45) do not depend on r we can represent the normal stresses in form

$$\begin{matrix}\sigma_r \\ \sigma_\theta\end{matrix} = \Omega(t)(A + r^{-2\mu}(K(\theta) \pm g^\mu \cos 2\psi)) \tag{5.50}$$

where according to the first equilibrium law (2.67)

$$K(\theta) = ((g^\mu \sin 2\psi)' + 2(1-\mu)g^\mu \cos 2\psi)/2\mu.$$

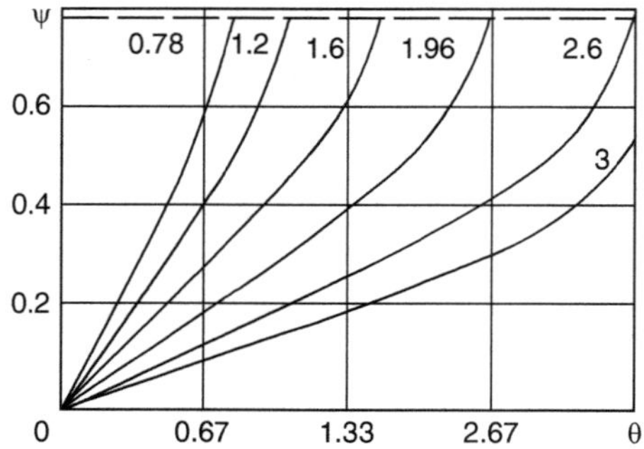

Fig. 5.9. Function $\psi(\theta)$ at different λ and $\mu = 0.5$

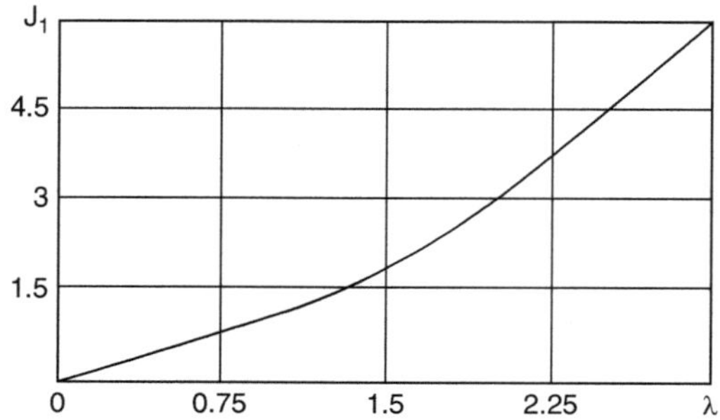

Fig. 5.10. Dependence of J on λ at $\mu = 0.5$

Using the first integral static equation (3.32) and condition $\sigma_\theta(a, \lambda) = \sigma_r(a, \lambda) = -q_*$ we have

$$\begin{matrix}\sigma_r \\ \sigma_\theta\end{matrix} = q_*(-1 + (g^\mu(\theta) \sin 2\psi)'/_{\theta=\lambda} + (a/r)^{2\mu} 2\mu(K(\theta) \pm g^\mu(\theta) \cos 2\psi))/B_3,$$

$$\tau = 2\mu(q_*/B_3)(a/r)^{2\mu} g^\mu(\theta) \sin 2\psi,\ \tau_e = 2\mu(q_*/B_3)(a/r)^{2\mu} g^\mu(\theta). \tag{5.51}$$

Here

$$B_3 = -(g^\mu(\theta) \sin 2\psi)'|_{\theta=\lambda} - (g^\mu(\lambda) \cos \lambda + J(\lambda))/\sin \lambda$$

and (see Fig. 5.10 for $\mu = 0.5$)

$$J = \int_0^\lambda g^\mu(\theta)\,(\sin 2\psi \sin \theta + 2 \cos 2\psi \cos \theta)\, d\theta.$$

The maximum τ_e is at $r = a$ and here we can use the criterion of infinite rate of the biggest elongation ε which gives with consideration of (5.28) and the second expression (5.51)

$$\varepsilon_* = \frac{1}{\alpha}, \Omega(t_*) = ((2\mu q_*/B_3) \max g^\mu(\theta) e\alpha)^{-m}. \tag{5.52}$$

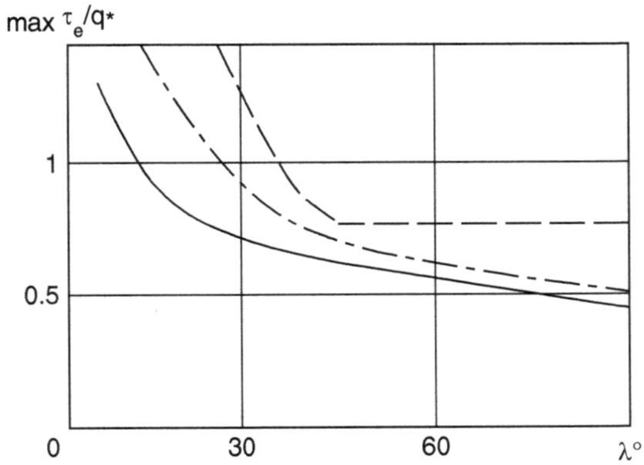

Fig. 5.11. Dependence max $\tau_e(\lambda)$ at different μ

Some Particular Cases

At $\mu = 0$ we have from (5.46) expression (4.14) and hence the solution of Sect. 4.2.2. So from (4.15) we find (solid line in Fig. 5.11)

$$\tau_e = \tau_{yi} = q_*/n\ln(n/(n-1)).$$

where n is linked with λ by relation in the above mentioned paragraph.

At $\mu = 1$ we derive from (5.46), (5.42) and the anti-symmetry condition

$$g \sin 2\psi = 2D \sin 2\theta \tag{5.53}$$

and from (5.51) we derive (broken line in Fig. 5.11)

$$\max \tau_e/q_* = 0.75x^{1}_{\cot\lambda} \quad \begin{matrix} \lambda \geq \pi/4 \\ \lambda \leq \pi/4 \end{matrix} \tag{5.54}$$

At $\mu = 0.5$ we have from (5.46)

$$\sqrt{g}\sin 2\psi = H \sin \theta \tag{5.55}$$

where H is a constant. Putting (5.55) into (5.42) we receive differential equation

$$d\theta/d\psi = (1 + 2\cot^2 2\psi)/(1 + \cot\theta \cot 2\psi). \tag{5.56}$$

which should be integrated at different $\Phi(0) = \Phi_o$ for boundary condition $\theta(0) = 0$ and it gives values of λ at $\psi = \pi/4$. Sokolovski /18/ has represented the results for $\lambda < \pi/4$. We made the computations for all $\lambda < \pi$ (Fig. 5.9). From (5.51), (5.55) and (2.65) we find maximum shearing stress

$$\tau_e = 2q_* \sin\theta/\cos 2\psi B_4. \tag{5.57}$$

Here

$$B_4 = (2J_1(\lambda) + \lambda)/\sin\lambda - \cos\lambda$$

and

$$J_1 = \int_0^{\lambda} (\sin 2\theta/\tan 2\psi)\, d\theta.$$

Seeking $d\tau_e/d\theta = 0$ we have with the consideration of (5.56) condition $\tan 2\psi = 2\tan\theta$ which gives to (5.57) at $\theta = \lambda, a = r$

$$\max \tau_e = q_* \sqrt{\cos^2\lambda + 4\sin^2\lambda}/B_4. \quad (5.58)$$

Diagram of (5.58) is drawn in Fig. 5.11 by broken-pointed curve.

5.2.4 Penetration of Wedge and Load-bearing Capacity of Piles Sheet

Putting σ_r, σ_θ from (5.50) and τ from (5.45) into (4.20) we receive

$$P/2l = p_*(1 + a/l)\sin\lambda + \omega B_5/a^{2\mu} \quad (5.59)$$

where

$$B_5 = ((1 + l/a)^{1-2\mu} - 1)(K(\lambda)\sin\lambda + g^{\mu}(\lambda)\cos\lambda(a/l)/(1-2\mu) - J_2(\lambda)),$$

$$J_2 = \int_0^{\lambda} ((K(\theta) + g^{\mu}\cos 2\psi)\cos\theta + g^{\mu}\sin 2\psi \sin\theta)\, d\theta.$$

Computing according to (2.65), (5.45) τ_e we find for its maximum at $r = a$

$$\max \tau_e = (P/2l - p_*(1 + a/l)) \max g^{\mu}(\theta)/B_5. \quad (5.60)$$

At $a \to 0$ we have the case of the wedge penetration and at $a \to \infty, \lambda \to \pi$ we come to the load-bearing capacity of piles sheet. Now we consider the particular cases.

If $\mu = 0$ we have the solution of Sect. 4.23. At $\mu = 1$ we compute from (2.65), (3.31) at $C = 0$ and (4.14)

$$\max \tau_e = 2\,(P/2l\sin\lambda - p_*\,(1 + a/l))\, x^{\sin\lambda}_{\cos\lambda}/\left(1 + (4/3)\sin^2\lambda\right)$$
$$- \left(3 - 4\sin^2\lambda\right)/(1 + l/a)\Big) \begin{matrix} \text{at } \pi/2 \le \lambda \le 3\pi/4 \\ \text{at } 3\pi/4 \le \lambda \le \pi \end{matrix}$$

which gives at $a = 0$ and $a \to \infty, \lambda \to \pi$ respectively

$$\max \tau_e = 2\,(P/2l\sin\lambda - p_*)\, x^{\sin\lambda}_{\cos\lambda}/(1 + (4/3)\sin^2\lambda, \begin{matrix} \text{at } \pi/2 \le \lambda \le 3\pi/4 \\ \text{at } 3\pi/4 \le \lambda \le \pi \end{matrix}, P_{yi}$$
$$= 2\,(p_* b + \tau_{yi} l) \quad (5.61)$$

where both members are obvious and the last member in the second expression is added from mechanical point of view.

At $\mu = 0.5$ we receive with (4.20), (5.50) and (5.55).

$$\begin{aligned}\max\tau_e &= (P/2l - p_* (1 + a/l)\sin\lambda)\sqrt{\cos^2\lambda + 4\sin^2\lambda}/2\,((a/l) \\ &\quad\times \ln(l/a + 1)\sin 2\lambda - (\lambda + J_1(\lambda)))\,.\end{aligned} \tag{5.62}$$

For the cases $a = 0$ and $a \to \infty, \lambda \to \pi$ we have respectively

$$\begin{aligned}\max\tau_e &= (P/2l - p_*\sin\lambda)\sqrt{\cos^2\lambda + 4\sin^2\lambda}/2\,(\lambda + 2J_1(\lambda))\,, \\ \max\tau_e &= (P/2 - p_* b)/2\,(\pi + 2J_1(\pi))\end{aligned} \tag{5.63}$$

and the expression like the second one (5.61) and similar to (5.29) as well as according to the criterion $d\gamma/dt \to \infty$

$$P_u = 2(16\tau_t l + p_* b);\, \varepsilon_* = 1/\alpha, \Omega(t_*) = (\alpha e 2\max\tau_e)^{-2} \tag{5.64}$$

Here τ_u is ultimate shearing stress and for the last relations the following constitutive equation is used (see rheological law (5.28) at $\mu = 1/2$.)

$$\Omega(t)(2\max\tau_e)^2 = \varepsilon e^{-\alpha\varepsilon}$$

5.2.5 Wedge Under Bending Moment in its Apex

We will seek a rigorous solution of this task (Fig. 5.12) at $\sigma_\theta = 0$. Then we have from expressions (2.67) at $\tau_{r\theta} \equiv \tau$ equations /18/

$$\partial(r\sigma_r)/\partial r + \partial\tau/\partial\theta = 0, \partial(r^2\tau)/\partial r = 0 \tag{5.65}$$

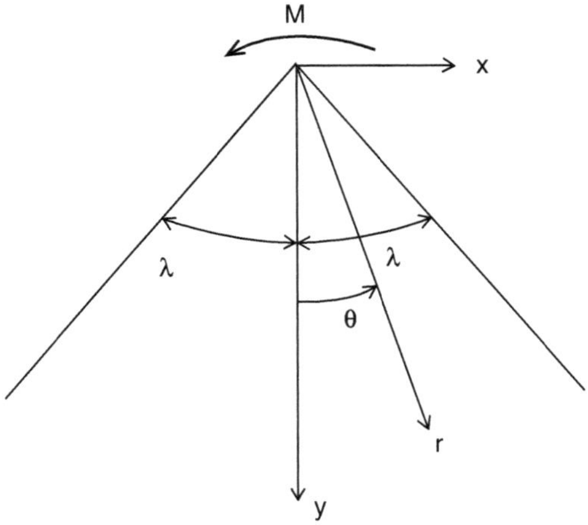

Fig. 5.12. Wedge under moment in its apex

that give solutions in form

$$\tau = f(\theta)/r^2, \sigma_r = f'(\theta)/r^2 \tag{5.66}$$

where f is a function of θ only. Boundary conditions for τ are

$$\tau(\pm\lambda) = 0. \tag{5.67}$$

Above that on the neutral axis ($\theta = 0$) as in the beam at bending similar to the problem of the slope under one-sided load we have $\sigma_r = 0$.

Components of main vector in any cross-section r = constant

$$\sum X = r \int_{-\lambda}^{\lambda} (\sigma_r \sin\theta + \tau\cos\theta)\, d\theta, \sum Y = r \int_{-\lambda}^{\lambda} (\sigma_r \cos\theta + \tau\sin\theta)\, d\theta$$

with a help of (5.65), (5.66) can be rewritten as

$$\int_{\lambda}^{\lambda} (f\sin\theta)'\, d\theta, \int_{\lambda}^{\lambda} (f\cos\theta)'\, d\theta$$

and according to (5.66), (5.67) they are equal to zero.

Now we use representations similar to (5.18)

$$\varepsilon_r = 0.5g(\theta)r^{-2m}\cos 2\psi, \gamma = g(\theta)r^{-2m}\sin 2\psi \tag{5.68}$$

where $g(\theta) = \gamma_m r^{2m}$ and according to (5.17)

$$\sigma_r = 2\omega(t)(g^{\mu}/r^2)\cos 2\psi, \tau = \omega(t)(g^{\mu}/r^2)\sin 2\psi. \tag{5.69}$$

Putting (5.69) into the first static equation (5.65) we find

$$(g^{\mu}\sin 2\psi)' - 2g^{\mu}\cos 2\psi = 0. \tag{5.70}$$

In the same manner we derive from (5.68), (2.71)

$$(g\cos 2\psi)'' + 4m(1-m)g\cos 2\psi + 2(2m-1)(g\sin 2\psi)' = 0. \tag{5.71}$$

At $\mu = m = 1$ we have from (5.70), (5.71) equation

$$(g\cos 2\psi)'' + 4g\cos 2\psi = 0$$

with obvious solution

$$g\cos 2\psi = C\cos 2\theta + D\sin 2\theta.$$

Condition above $\sigma_r(r,0) = 0$ gives $C = 0$ and from (5.65), (5.66) as well as obvious integral static law

$$M = -2\int_0^{\lambda} \tau(\theta)\, r^2 d\theta \tag{5.72}$$

we compute

$$\sigma_r = 2(M/r^2B_6)\sin 2\theta, \tau = (M/r^2B_6)(\cos 2\lambda - \cos 2\theta) \tag{5.73}$$

where

$$B_6 = \sin 2\lambda - 2\lambda\cos 2\lambda. \tag{5.74}$$

Now (2.65) gives

$$\tau_e = M(1+\cos^2 2\lambda - 2\cos 2\theta\cos 2\lambda)^{0.5}/r^2B_6. \tag{5.75}$$

Condition $d\tau_e/d\theta = 0$ leads to equality $\sin 2\theta = 0$ with solutions $\theta = 0$ and $\theta = \pi/2$. Investigations show that $\max\tau_e$ is at $\theta = 0$ in form

$$\max\tau_e = 2M\sin^2\lambda/B_6. \tag{5.76}$$

Diagram $\max\tau_e(\lambda)$ is drawn by solid line in Fig. 5.13. From relations (5.60), (5.73) we derive expression (5.22) with sign minus and hence, solid lines in Fig. 5.8 reflected relatively to axis θ.

In the general case we derive from (5.70), (5.71) after an exclusion of g the second order differential equation

$$\tan 2\psi\psi'' - 2(1/\mu - 1 + 2/\Psi)\psi'^2 + 2(1+2/\mu\Psi)\psi' - 2/\mu\Psi = 0 \tag{5.77}$$

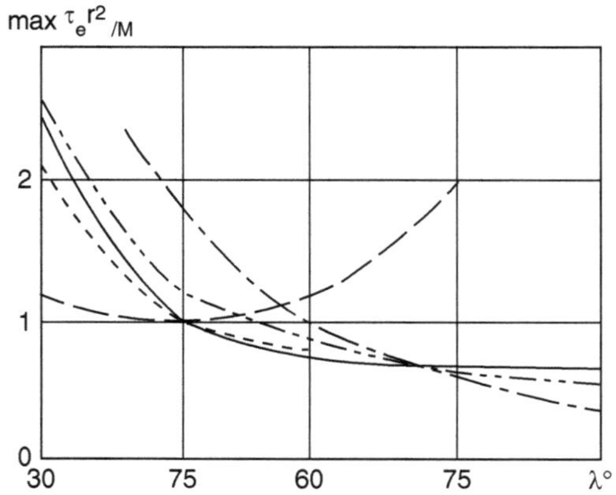

Fig. 5.13. Diagram $\max\tau_e(\lambda)$

where Ψ is given by relation (5.49). Now we suppose

$$d\theta/d\psi = \Theta \tag{5.78}$$

and (5.77) becomes

$$(\tan 2\psi)d\Theta/\Theta d\psi + 2(1/\mu - 1 + 2/\Psi) - 2(1 + 2/\mu\Psi)\Theta + 2\Theta^2/\mu\Psi = 0. \tag{5.79}$$

According to (5.7) equation (5.79) must be solved at different $\Theta(-\pi/4) = \Theta_o$. Then we integrate (5.78) with border demand $\theta(-\pi/4) = 0$ (see Appendix G).

To find $\max \tau_e$ we write from (5.70) with consideration of (5.78) and condition $g(0) = 1$ as

$$g(\theta) = \exp\left((2/\mu)\int_0^\theta (1 - 1/\Theta)\cot 2\psi d\theta\right). \tag{5.80}$$

Now from (5.69), (5.72) we find at $\theta = 0$ for $\lambda > \pi/4$ and $\theta = \lambda$ for $\lambda < \pi/4$ respectively

$$\max \tau_e = M/2Jr^2, \max \tau_e = Mg(\lambda)/2Jr^2. \tag{5.81}$$

V. Sokolovski /18/ used the first of them for $\lambda < \pi/4$ that underestimates $\max \tau_e$. The same mistake is in /18/ for a wedge under one-sided load (see Sect. 5.2.2). In (5.81)

$$J = \int_0^\lambda (g(\theta)/g(0))^\mu \sin 2\psi d\theta. \tag{5.82}$$

A very simple solution takes place at $\mu = 0$ when we have from (5.70) with consideration of boundary condition above relation

$$\psi = \theta - \pi/4$$

and from (5.69) we find expression

$$\tau = -(\omega(t)/r^2)\cos 2\theta.$$

Putting it into (5.72) we receive at $\theta = 0$ (broken line in Fig. 5.13)

$$\max \tau_e = M/r^2 \sin 2\lambda. \tag{5.83}$$

We can see that this value coincides with (5.78) only at $\lambda = \pi/4$ and at bigger values of λ it is above the curve for $m = 1$.

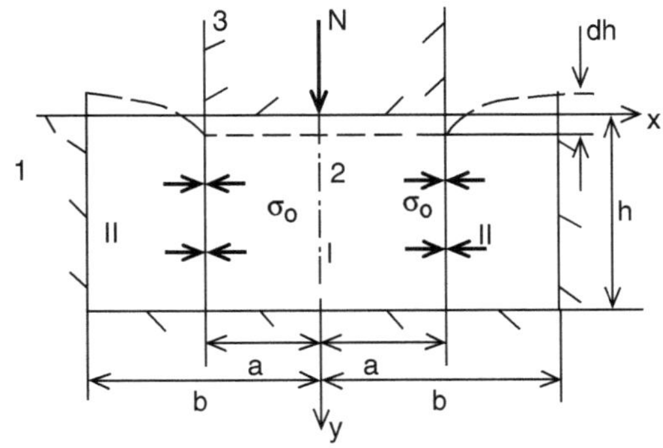

Fig. 5.14. Sliding support

5.2.6 Load-bearing Capacity of Sliding Supports

In order to improve conditions of sluice and lock exploitation the inconvenient wheels and rollers are often replaced by sliding supports. The latter usually consists /26/ of embraced in metal holder 1 (Fig. 5.14) polymer skid 2 along which steel rail 3 moves. The structure hinders from longitudinal displacement of the skid and because of that the greatest interest has an appreciation of its resistance to compression.

A rigorous solution of this task with consideration of complex boundary conditions and peculiarities of the polymer's mechanical behaviour is hardly possible. At the same time constantly widening use of such structures in different branches of industry compels to seek simple and convenient for practice approaches. Because of that we give here the approximate solution based on results of study of this problem for simple media and some kinematic hypotheses.

Here we use the first expression (5.17) as

$$\sigma_1 - \sigma_3 = 4\omega(t)\varepsilon^{\mu}. \tag{5.84}$$

On the base of applied plasticity theory /10/ (see also Sect. 4.2.2) and with consideration of the Coulomb's friction law /11,27/ the following expressions for parts I and II of the skid (Fig. 5.14) are formulated (see Appendix H)

$$\begin{aligned} 4\omega\,(t)\,(\varepsilon_{xI})^{\mu} &= \sigma_o + /N/f/h\left(e^{2fa/h} - 1\right), \\ 4\omega\,(t)\,(\varepsilon_{yII})^{\mu} &= -fh\sigma_o/\left(e^{fh/(b-a)} - 1\right)(b-a). \end{aligned} \tag{5.85}$$

Here f is friction coefficient, dimensions a, b, h are shown in Fig. 5.14 and value of σ_o in the second expression (5.85) is determined from integral static equation. Excluding from (5.85) σ_o and using the constant volume demand in form

$$\varepsilon_{xI} = (\beta - 1)\varepsilon_{yII} \tag{5.86}$$

where $\beta = b/a$ we derive

$$4\omega(t)(\varepsilon_{xI})^{\mu} = /N/f/h(e^{2fh/a} - 1)(1 + (b-a)(e^{fh/(b-a)} - 1)/fh(\beta - 1)^{\mu}). \quad (5.87)$$

When ε_{xI} is known the most interesting value of track's depth dh can be found from (5.86) as

$$dh = h\varepsilon_{xI}\beta/(\beta - 1). \quad (5.88)$$

The approximate solution (5.87), (5.88) must be compared with similar results for ideal bodies according to the Hookè's law when $\mu = 1, \omega = G$, and perfect plasticity with $\mu = 0, \sigma_1 - \sigma_3 = \sigma_{yi}$. The comparison will be made for smooth surfaces contact of the skid with the holder and the rail. So, at $f = 0$ we have from (5.87), (5.88)

$$Gdh/N = h\beta/8b$$

(solid lines 1, 2 in Fig. 5.15 for $b/h = 1, 2$ respectively) and

$$a\sigma_{yi}/N = 1/4$$

(solid straight line in Fig. 5.16).

The solution of the problem for linear material is given in the form (broken line in Fig. 5.15)

$$dh = C - /N/F(\varphi, dn(K/\beta))/4GK_1 \quad (5.89)$$

where C-constant, F – incomplete elliptic integral of the first kind with modulus $dn(K'/\beta)$

$$\varphi = \sin^{-1}(sn(Kx/b)/sn(K/\beta)),$$

sn, dn – sine and delta of amplitude (the Jakobi's elliptic functions), K, K_1 – full elliptic integrals of the first kind with moduli k and $ksn(K/\beta)$ respectively. The dimensions are linked by relation

$$K(k)/K(k') = b/h. \quad (5.90)$$

Here $k' = \sqrt{1 - k^2}$ – modulus additional to k.

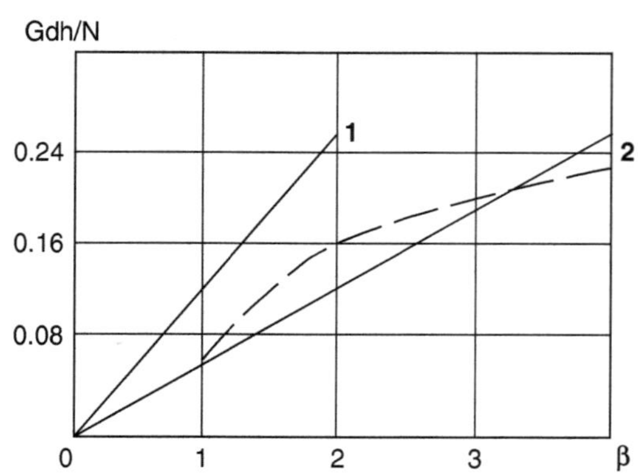

Fig. 5.15. Dependence of dh on β for elastic material

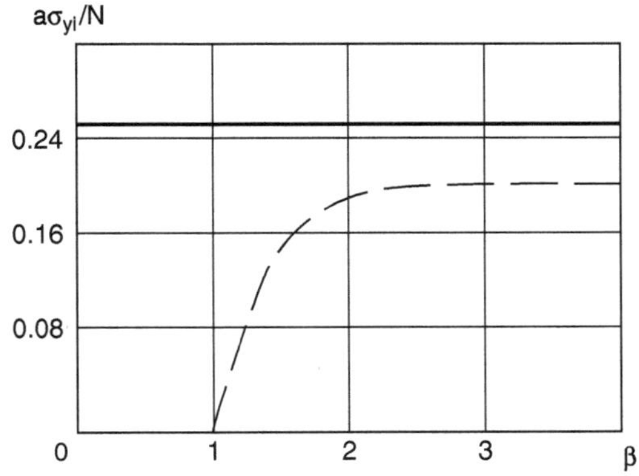

Fig. 5.16. Compression of ideal plastic layer

On the base of theoretical-experimental investigation R. Hill /28/ gave the dependence of punch pressure on the thickness of a layer of ideal plastic material that in terms of our work is drawn by broken line in Fig. 5.16. At $\beta > 2$ it is near to that of expressions (4.39).

From Figs. 5.15, 5.16 we can see that at b/h near 2 and $\beta > 2$ the proposed here simple solution agrees well enough with similar results for ideal bodies. Since in practice dimensions of the skid are taken as b near to 2 h and $\beta = 1.5$–3 the represented relations can be used for applications. To receive calculation expressions we exclude from (5.87). (5.88) ε_{xI} that gives

$$dh = \Omega(t)(N/4)^m R \tag{5.91}$$

where

$$R = \beta h^{1-m}/(f(e^{2fa/h} - 1)(1 + (b - a)(e^{fh/(b-a)} - 1)/fh(\beta - 1)^{\mu})^m(\beta - 1)).$$

Relation (5.91) has the same properties as similar creep laws for tension and in complex stress state and it can serve for computation of dh according to the test data in compression.

5.2.7 Propagation of Cracks and Plastic Zones near Punch Edges

General Relations

Similarly Sect. 5.1.1 we take the potential function as

$$\Phi = Kr^s f(\theta)$$

We find the stresses according to (2.75) and strains by (2.66) at $\alpha = 0$. Since sum of products $\sigma_{ij}\varepsilon_{ij}$ is proportional to r^{-1} /21/ we have s =

$(2m+1)/(m+1)$,

$$\sigma_r = Kr^{-1/(m+1)}(sf+f''), \sigma_\theta = Kr^{-1/(m+1)}s(s-1)f,$$
$$\tau_{r\theta} = Kr^{-1/(m+1)}(1-s)f' \qquad (5.92)$$

and

$$\varepsilon_\theta = -0.5\Omega_1(t)K^m r^{-m/(m+1)}F^{m-1}(f''+s(2-s)f),$$
$$\gamma_{r\theta} = 2\Omega_1(t)(1-s)K^m r^{-m/(m+1)}F^{m-1}f'. \qquad (5.93)$$

Here Ω_1 is proportional to Ω, x, y in (2.61) must be replaced by r, θ respectively and

$$F = ((f''+f(1+2m)/(m+1)^2)^2 + (2mf'/(m+1))^2)^{0.5}. \qquad (5.94)$$

Putting $\varepsilon_\theta, \gamma_{r\theta}$ into compatibility law (2.71) we get a very complex non-linear differential equation of the fourth order which must be solved together with the consequent boundary conditions.

Crack in Tension and Pressure of Punch

The border demands for these problems are respectively: $f(0) = 1, f'(0) = f'(\pi) = f(\pi) = 0$ and $f(\pi) = -1, f'(0) = f'(\pi) = f(0) = 0$.

Since for the tasks $f'(0) = f'(\pi) = 0$, in perfect plastic solution $f'''(0), f'''(\pi)$ and this value at $\pi/4 < \theta < 3\pi/4$ is also equal to zero as well as in consequent elastic relations (Sects. 3.2.7 and 3.2.9 where $f(\theta)$ is proportional to $\cos^3\theta$ and $\sin^3\theta$ respectively) $f'''(0) = 0$ we suppose $f' = f''' = 0$ everywhere in the zones mentioned above and derive equation (see also Appendix I)

$$f^{Iv} + (3m+7)f''/(1+m)^2 + (2m+1)(2+m)f/(m+1)^4 = 0 \qquad (5.95)$$

with obvious solution

$$f = A_1\cos\beta_1\theta + A_2\sin\beta_1\theta + A_3\cos\beta_2\theta + A_4\sin\beta_2\theta. \qquad (5.96)$$

Here

$$\begin{matrix}\beta_1\\ \beta_2\end{matrix} = \left(\left(-3m-7\pm\sqrt{m^2+22m+41}\right)/2(1+m)^2\right)^{0.5},$$

and A_j (j = 1–4) can be found from the boundary conditions (it is interesting to notice that at m = 1 this approximate solution coincides with the rigorous one – see consequent relations in Sect. 3.2.8). Then we compute f', f'' and the stresses according to (5.92). Calculations show that their diagrams for $m = 3, m = 15$ are near to ones in Fig. 4.19 for elastic and perfect plastic bodies respectively.

In the general case we can find K according to the condition /21/ of the independence of integral $J = dW/dl$ on the properties of the material. Here

$$dW = \int_0^{dl} \sigma_\theta(x, 0)\, u_\theta(dl - x, \pi)\, dx. \tag{5.97}$$

Putting σ_θ, u_θ from (3.90), (3.91) into (5.97) we find for $m = 1$ and $k = 1$ (plane deformation)

$$J_1 = \pi\sigma^2 l/4G. \tag{5.98}$$

To get J in the general case we use relations (5.93), (5.92) for ε_r, ε_θ, $\gamma_{r\theta}$, σ_θ and compute u_θ from (2.69). The latter gives

$$u_\theta = \Omega_1(t)K^m r^{1/(m+1)} F^{m-1}((6m+1)f' - (m+1)^2((m-1) \\ \times F'(f'' + fs(2-s))/F + f'''))/4m$$

where

$$F'/F = F^{-2}(f'' + fs(2-s))(f''' + f's(2-s)) + (4(1-s)f'f'').$$

Now from (5.97), (5.92) and expression for u_θ above we derive J for any m

$$J = 0.5K^{m+1}m(2m+1)/f''(\pi)/^{m-1}f'''(\pi) \int_0^1 ((1-\xi)/\xi)^{1/(1+m)}\, d\xi.$$

Using equation $J = J_1$ we have

$$K = ((\kappa+1)\pi\sigma^2/4G\Omega_1(t)I_1(m))^{1/(1+m)}.$$

Here

$$I_1(m) = (2m+1)(m+1)/f''(\pi)/^{m-1}f'''(\pi)\Gamma((2m+1)/(m+1))\Gamma((2+m)/(1+m))$$

and $\Gamma()$ is a Γ-function. Lastly, according to condition τ_e = constant we compute equation

$$3r(2\tau_e)^{m+1}G\Omega_1 I_1(m)/\sigma^2\pi l = F^{m+1} \tag{5.99}$$

which determines yielding zone near the crack edges.

Transversal Shear

It can be considered in a similar manner. Since in this problem $f(0) = f(\pi) = 0$ we neglect here f and f'' versus f' and find from (5.94)

$$F = 2mf'/(m+1).$$

Compatibility law (2.71) gives equation

$$f^{Iv} + (5m^2 + 4m + 1)f''/(m + 1)^2 + m(2m + 1)(m + 2)f/(m + 1)^4 = 0$$

with solution (5.96) in which

$$\begin{matrix}\beta_1 \\ \beta_2\end{matrix} = \left(\left(-(5m^2 + 4m + 1) \pm \sqrt{25m^4 + 32m^3 + 6m^2 + 1}\right)/2\right)^{0.5}/(m + 1).$$

For the crack and punch we have two other border demands as $f'(\pi) = 0$, $f'(0) = 1$ and $f'(\pi) = 1, f'(0) = 0$ respectively. It is interesting to notice that once again the approximate solution gives at $m = 1$ the rigorous results.

Now we find f', f'' (see calculations in Appendix J) and according to (5.92) – stresses σ_θ, σ_r, $\tau_{r\theta}$, τ_e, strain ε_θ, displacement u_r, integral J and factor K as

$$K^{m+1} = 3\pi\tau^2 l/I(m)/f''(\pi)/^m$$

where I(m) is the same as in Sect. 5.1.2. Condition τ_e = constant gives equation for r in form

$$1.5r(2\tau_e)^{m+1}G\Omega(t)/\tau^2\pi l = F^{m+1}/I(m)/f''(\pi)/^m. \tag{5.100}$$

Diagrams $\sigma_\theta/\sigma_{yi}, \sigma_r/\sigma_{yi}, \tau_{r\theta}/\sigma_{yi}$ are given in Fig. 3.21 by the same lines as for $m = 1$ but with index 0. From the figure we can see that with the growth of m the distribution of stresses changes very strongly. In the same manner the problem of the punch horizontal movement can be considered. The curves for the stresses can be received by reflection of the previous ones relatively to axis $\theta = \pi/2$.

5.3 Axisymmetric Problem

5.3.1 Generalization of Boussinesq's Solution

As in Sect. 3.3.2 we suppose for incompressible material ($\nu = 0.5$) $\sigma_\chi = \sigma_\theta = \tau_{\rho\chi} = 0$, and from the first static equation (2.77) as well as from rheological law (1.45) at $\alpha = 0$ we have for stress and strain following relations /31/

$$\sigma_\rho = f(\chi)/\rho^2, \varepsilon_\rho = g(\chi)\Omega(t)\rho^{-2m} \tag{5.101}$$

where $g = f^m$. Since for this case in (2.79) $\varepsilon_\chi = \varepsilon_\theta$ we find easily

$$u_\chi = U(\rho)\sin\chi, u_\rho = \varphi(\chi) + g(\chi)\Omega(t)\rho^{1-2m}/(1 - 2m). \tag{5.102}$$

Using condition $\varepsilon_\rho = -2\varepsilon_\chi$ we derive

$$0.5\Omega(t)(3 - 2m)g(\chi)\rho^{1-2m}/(1 - 2m) + \varphi(\chi) + U(\rho)\cos\chi = 0. \tag{5.103}$$

Putting u_ρ, u_χ from (5.102) into condition $\gamma_{\rho\chi} = 0$ we determine

$$\varphi'(\chi) + g'(\chi)\Omega(t)\rho^{1-2m}/(1-2m) + \rho^2(\sin\chi)\partial(U(\rho)/\rho)/\partial\rho = 0. \quad (5.104)$$

Excluding $\varphi'(\chi)$ from (5.103), (5.104) we obtain the expression in which both parts must be equal to the same constant, say n, since each of them depends only on one variable (neglecting t as a parameter) in form

$$\Omega(t)g'(\chi)/2\sin\chi = \rho^{2m}dU(\rho)/d\rho = -n$$

with obvious solutions

$$f(\chi) = \Omega^{-\mu}(C + 2n\cos\chi)^{\mu}, U(\rho) = D - n\rho^{1-2m}/(1-2m). \quad (5.105)$$

Since at $\chi = \pi/2$ we have $\sigma_\rho = 0$ we must put in the first (5.105) $C = 0$ and constant n should be found from condition (Fig. 5.17)

$$P = -2\int_0^{\lambda} \sigma_\rho \rho^2 \sin\chi\cos\chi d\chi. \quad (5.106)$$

Putting here σ_ρ from (5.101) we find after calculations

$$\sigma_\rho = -P(\mu+2)(\cos\chi)^{\mu}/2\pi(1-\cos^{\mu+2}\lambda)\rho^2. \quad (5.107)$$

Taking in the second relation (5.105) $D = 0$ we get the displacement as

$$u_\chi = -0.5\Omega(t)(P/(\mu+2)/2\pi(1-\cos^{2+\mu}\lambda))^m\rho^{1-2m}(\sin\chi)/(1-2m). \quad (5.108)$$

The most interesting case takes place at $\lambda = \pi/2$ when we receive from expressions (5.107), (5.108)

$$\begin{aligned} \sigma_\rho &= -P(2+\mu)\left(\cos^{\mu}\chi\right)/2\pi\rho^2, \\ u_\chi &= -0.5\Omega\,(t)\,(P(2+\mu)/2\pi)^m\rho^{1-2m}\,(\sin\chi)\,/(1-2m). \end{aligned} \quad (5.109)$$

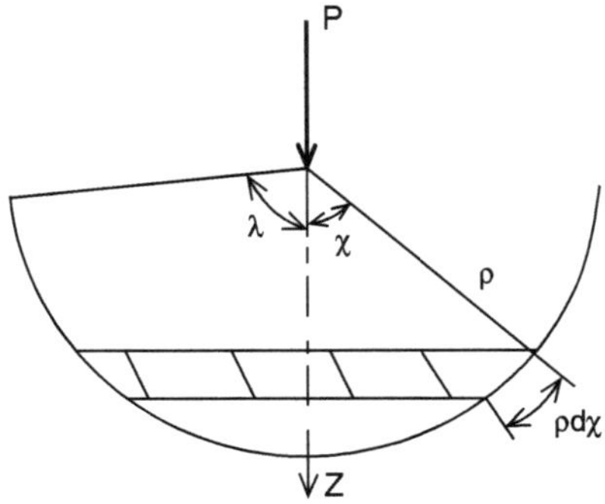

Fig. 5.17. Computation of constant n

It is easy to notice that the highest value of σ_ρ at ρ = constant is on the line $\chi = 0$. It is not difficult to find that there stress σ_ρ at m = 1 is 1.5 times more than at $\mu = 0$. The biggest value of u_χ is at $\chi = \pi/2$ but its dependence on m is more complex. However the second relation (5.109) allows to calculate the displacements in some distance from the structure loaded by forces with a resultant P.

To appreciate a practical meaning of the results we compare for m = 1 the distribution of stress σ_z on axis z for the concentrated force $P = q\pi a^2$ and for the circular punch of radius a when we have from (5.109) and (3.122) respectively

$$/\sigma_z//q = 3(a/z)^2/2, \sigma_z/q = 1 - (1 + (a/z)^2)^{-3/2}. \tag{5.110}$$

From Fig. 5.18 where by solid and broken lines diagrams $/\sigma_z/(z)$ are shown we can see that at $z/a > 3$ the simplest solution for concentrated force can be used. Since at $\mu < 1$ a distribution of stresses becomes more even we can expect better coincidence of similar curves with the growth of a non-linearity

It is interesting to notice that according to Fig. 5.18, 5.4 vertical stress in axisymmetric problem is approximately twice less than in the plane one. This explains higher load-bearing capacity of compact foundations.

5.3.2 Flow of Material within Cone

Common Equations

We solve this problem at the same suppositions as that in Sect. 4.3.3 from (2.79) at $u_\chi = 0$ we compute

$$\begin{aligned} &\varepsilon_\rho = -2U/\rho^3, \varepsilon_\theta = \varepsilon_\chi = U/\rho^3, \\ &\gamma_{\rho\chi} \equiv \gamma = dU/\rho^3 d\chi, \gamma_m = g/\rho^3 \end{aligned} \tag{5.111}$$

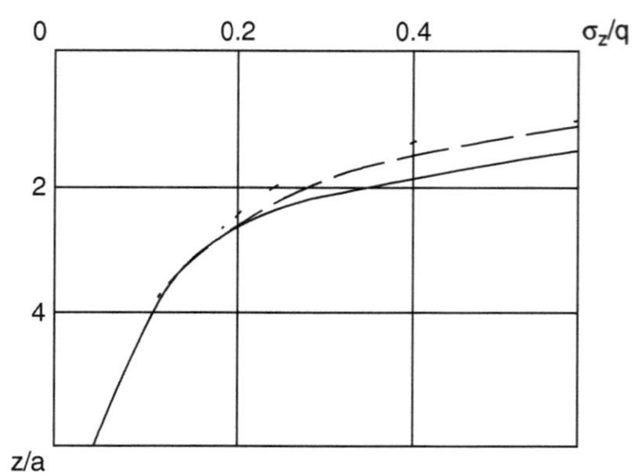

Fig. 5.18. Comparison of stress distribution

where $U = U(\chi)$ and according to (2.16)

$$g(\chi) = \sqrt{9U^2 + U'^2}.$$

Similar to (5.18) and (5.68) we use representations

$$\varepsilon_\chi = -g(\cos 2\psi)/3\rho^3, \gamma_{\rho\chi} = g(\sin 2\psi)/\rho^3 \tag{5.112}$$

putting which into the first law (2.82) we have equations

$$\begin{aligned} (g\cos 2\psi)' + 3g\sin 2\psi = 0, \\ dg/gd\chi = 2\,(d\psi/d\chi - 3/2)\tan 2\psi. \end{aligned} \tag{5.113}$$

The latter gives boundary condition $d\psi/d\chi = 3/2$ at $\psi = \pi/4$. From expressions for strains above we can also find

$$d\ln/U//d\chi = -3\tan 2\psi, g(\chi) = -3U/\cos 2\psi, \tag{5.114}$$

From (5.17) and (5.112) we derive representations

$$\begin{aligned} \tau_{\rho\chi} = \tau = \omega\,(t)\,\rho^{-3\mu} g^{\mu} \sin 2\psi, \\ \begin{matrix}\sigma_\rho \\ \sigma_\chi\end{matrix} = \omega\,(t)\,(C + \rho^{-3\mu}\,(K^{+2}_{-1} x 2g^{\mu}\,(\cos 2\psi)/3)) \end{aligned} \tag{5.115}$$

where C is a constant and function $K(\theta)$ can be found from the first static equation (2.77) as follows

$$3\mu K = (g^\mu \sin 2\psi)' + \cot\chi(g^\mu \sin 2\psi) + 4(1-\mu)g^\mu \cos 2\psi. \tag{5.116}$$

Putting (3.115) into (2.78) we derive

$$\begin{aligned} (g^\mu \sin 2\psi)'' + (g^\mu \sin 2\psi)'\cot\chi + (9\mu(1-\mu) - 1/\sin^2\chi)g^\mu \sin 2\psi \\ + 2(2-3\mu)(g^\mu \cos 2\psi)' = 0. \end{aligned} \tag{5.117}$$

Combining (5.113), (5.117) we have at Ψ according to (5.49) two differential equations

$$\Theta = d\chi/d\psi, \tag{5.118}$$

$$\begin{aligned} (\cot 2\psi)\,d\Theta/d\chi - 2(\mu - 1 + 2\mu/\Psi) + \Theta((6\mu^2 + 3\mu + 4\,(1-\mu)\cos^2 2\psi)/\Psi \\ - \cot\chi\cot 2\psi) - 3\Theta^2(3\mu^2 - (\mu + \mu\tan 2\psi\cot\chi + 1/3\sin^2\chi)\cos^2 2\psi)/2\Psi = 0 \end{aligned} \tag{5.119}$$

the second of which should be solved at different $\Theta_o = \Theta$. Then we integrate (3.118) at border demand $\chi(0) = 0$. The searched function must also satisfy condition $\chi = \lambda$ at $\psi = \pi/4$. Now we receive from (5.113), (5.114), (2.65) $U(\chi), g(\chi)$ and τ_e.

Putting stress σ_ρ from (5.115) into integral static equation (4.105) we find at $-q_* = \sigma_\rho(a,\lambda) = \sigma_\chi(a,\lambda)$ expression for $\max\tau_e$ as

$$\max\tau_e = 3\mu q_* \max g^\mu(\chi)/((g^\mu(\theta)\sin 2\psi)'|_{\chi=\lambda} - g^\mu(\lambda)\cot\lambda - 2J_3/\sin^2\lambda)$$

where

$$J_3 = \int_0^\lambda g^\mu(\theta)(\sin 2\psi \sin^2\chi + 2\cos 2\psi \sin 2\chi)d\chi.$$

Then the criteria $\max\tau_e = \tau_u$ and $d\gamma_m/dt \to \infty$ must be used as before. For the latter we have

$$\varepsilon_* = 1/\alpha, \Omega(t_*) = (\alpha e 2 \max\tau_e)^{-m}.$$

Some Particular Cases

At $\mu = 0$ we have from (5.117), (5.103) and hence the solution of Sect. 4.3.3. If $\mu = 1$ we compute from (5.113), (5.117) equation

$$(g\sin 2\psi)'' + (g\sin 2\psi)'\cot\chi + (6 - 1/\sin^2\chi)g\sin 2\psi = 0$$

with obvious solution

$$g\sin 2\psi = 2D\sin 2\chi \tag{5.120}$$

where D is a constant. Then from (5.112) and the stick condition

$$g\cos 2\psi = 3D(\cos 2\chi - \cos 2\lambda). \tag{5.121}$$

From (5.120), (5.121) we receive

$$\tan 2\psi = 2(\sin 2\chi)/3(\cos 2\chi - \cos 2\lambda)$$

Diagrams $\psi(\chi)$ at different λ according to this relation are drawn in Fig. 5.19. Similar to the general case we have ultimate condition as

$$\max\tau_e = q_* x^1_{2/3\tan\lambda}(\lambda \gtrless 33.7^\circ) \tag{5.122}$$

At $\mu = 2/3$ we calculate from (5.117) equation

$$(g^{2/3}\sin 2\psi)'' + (g^{2/3}\sin 2\psi)'\cot\chi + (2 - 1/\sin^2\chi)g^{2/3}\sin 2\psi = 0$$

with obvious solution

$$g^{2/3}\sin 2\psi = H\sin\chi \tag{5.123}$$

where H is a constant. Putting (5.123) into (5.113) we derive differential equation of the first order

$$d\chi/d\psi = 2(2 + 3\cot^2 2\psi)/3(2 + \cot\chi\cot 2\psi) \tag{5.124}$$

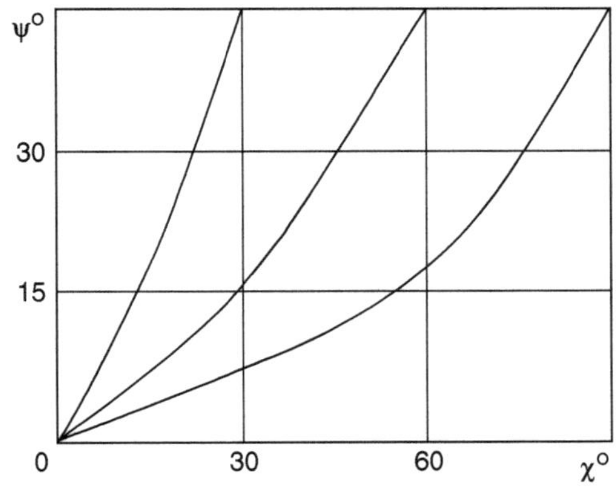

Fig. 5.19. Dependence $\psi(\chi)$ at $\mu = 1$

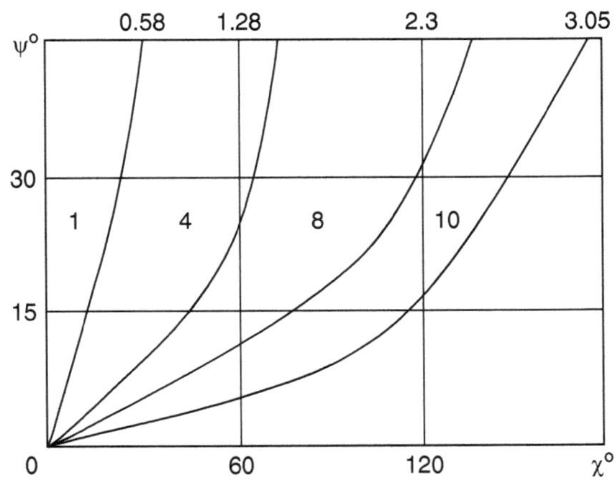

Fig. 5.20. Diagrams $\psi(\chi)$ at $\mu = 2/3$ and different Θ_o

that should be integrated at different $\Theta(0) = \Theta_o$. Diagrams $\chi(\psi)$ at Θ_o-values in the curves middles and λ at their tops are given in Fig. 5.20.

Putting σ_ρ from (5.113) into (4.105) we find C and from condition $d\tau_e/d\chi = 0$ with consideration of (5.124) – equality $\tan 2\psi = 3\tan\chi$ which gives to $\max\tau_e$ (it increases with a growth of χ) value

$$\max\tau_e = q_*\sin^2\lambda\sqrt{\cos^2\lambda + 9\sin^2\lambda}/\left(3\cos\lambda - \cos^3\lambda - 2 - 12J_4\right).$$

Here as before $-q_* = \sigma_\rho(a,\lambda) = \sigma_\chi(a,\lambda)$ and

$$J_4 = \int_0^\lambda (\sin^2\chi\cos\chi)\,(\tan 2\psi)^{-1}\,d\chi.$$

Diagrams $J_4(\lambda)$ and $\max\tau_e(\lambda)$ are shown by solid curves in Figs. 5.21, 5.22 respectively. The broken line in the latter picture refers to the case $\mu = 1$

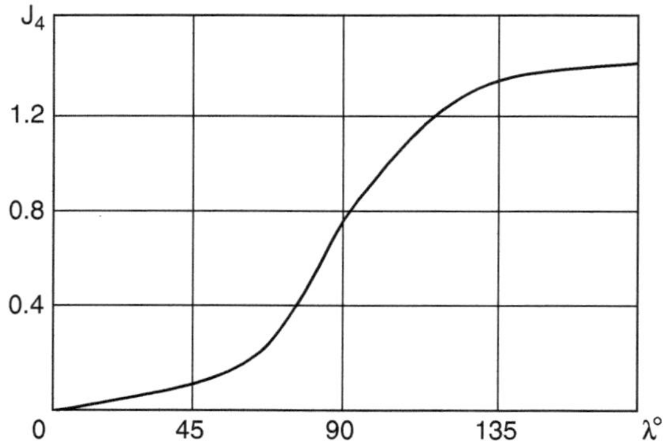

Fig. 5.21. Diagram $J_4(\lambda)$ for $\mu = 2/3$

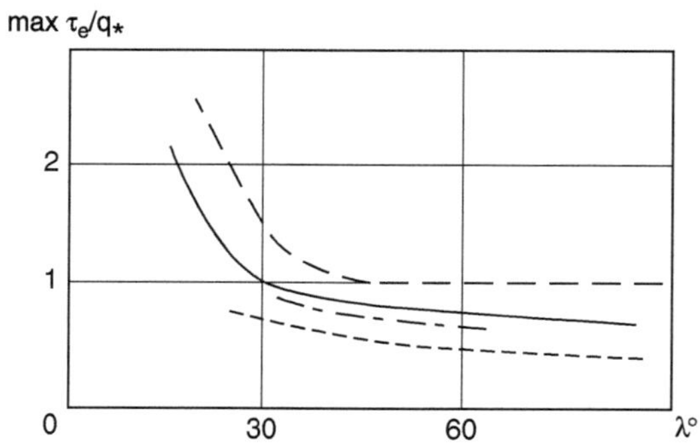

Fig. 5.22. Diagrams $\max \tau_e(\lambda)$

(computations for $\mu = 1/3$ see in Appendix K interrupted by points curve in the figure) and pointed line refers to solution (4.106).

5.3.3 Cone Penetration and Load-bearing Capacity of Circular Pile

Here common relations (5.111)–(5.119) are valid. We put stresses according to (5.115) into integral static equations (4.107) to detail the constants and according to (2.65) we compute $\max \tau_e$ at $a = \rho$ as

$$\begin{aligned}\max \tau_e = {} & 3\mu(P/\pi - p_*(a+1)^2 \sin^2\lambda)(\max g^\mu(\chi))/a(a(2(g^\mu(\chi)\sin 2\psi)'|_{\chi=\lambda} \\ & \times \sin^2\lambda + g^\mu(\lambda)(1+3\mu)\sin 2\lambda(1+l/a)^{2-3\mu} - 1)/(2-3\mu) \\ & - l(g^\mu(\lambda)\sin 2\lambda + 2J_5)(2+l/a)) \qquad (5.125)\end{aligned}$$

where p_* is the strength of soil in a massif at compression and

$$J_5 = \int_0^\lambda g^\mu(\chi)((1+3\mu)\sin 2\psi \sin^2\chi + 2\cos 2\psi \sin 2\chi)d\chi.$$

At $\lambda \to \pi, a \to \infty$ we find for a circular pile

$$\max \tau_e = (P/\pi - p_* b^2) \max g^{\mu}(\chi)/4l(bg^{\mu}(\lambda) + J_5(\lambda))(2 + l/a). \qquad (5.126)$$

In the same manner we consider the particular cases and consequently for $\mu = 0$ we receive from (5.117), (4.103) and hence the solution of Sect. 4.3.4.

At $\mu = 1$ we have

$$\max \tau_e = 4.5(P/\pi - p_*(a+l)^2 \sin^2\lambda) x^1_{2/3 \tan\lambda}/la(2(5 - 6\sin^2\lambda)/(1 + l/a)$$
$$-(2 + 3\sin^2\lambda)(2 + l/a))^{\lambda<146^\circ}_{\lambda>146^\circ}$$

and for the pile the yielding (the first ultimate) load is

$$P_{yi} = \pi b(2\tau_{yi} l + p_* b). \qquad (5.127)$$

We can see that this result has obvious structure and coincides with approximate relation (4.110) for ideal plasticity.

Similarly we compute for $\mu = 2/3$

$$\max \tau_e = (P/\pi - p_*(a+l)^2 \sin^2\lambda)\sqrt{\cos^2\lambda + 9\sin^2\lambda}/6a(2a\cos\lambda \sin^2\lambda$$
$$\times \ln(1 + l/a) - l(2 + l/a)(1 - \cos\lambda + 2J_4)).$$

Here J_4 is given in Sect. 5.3.2. For the circular pile this relation predicts big values of ultimate load and so we can take in the safety side

$$P_u = \pi b(p_* b + 2\tau_u l). \qquad (5.128)$$

5.3.4 Fracture of Thick-walled Elements due to Damage

Stretched Plate with Hole

We consider plate of thickness h with axes r, θ, z (Fig. 5.23) and use the Tresca-Saint-Venant hypotheses. Since here $\sigma_\theta > \sigma_r > \sigma_z = 0$ we have $\varepsilon_r = 0$ and from (2.32) at $\alpha = 0, \sigma_{eq} = 2\tau_e = \sigma_\theta$

$$\varepsilon \equiv \varepsilon_\theta = 3\Omega(t)\sigma_\theta^m/4$$

$$\sigma_\theta = (4/3\Omega)^{\mu}\varepsilon^{\mu} \qquad (5.129)$$

where $\varepsilon = u/r$ and radial displacement u depends only on t.

Putting (5.129) into the static equation of this task

$$h\sigma_\theta = d(hr\sigma_r)/dr, \qquad (5.130)$$

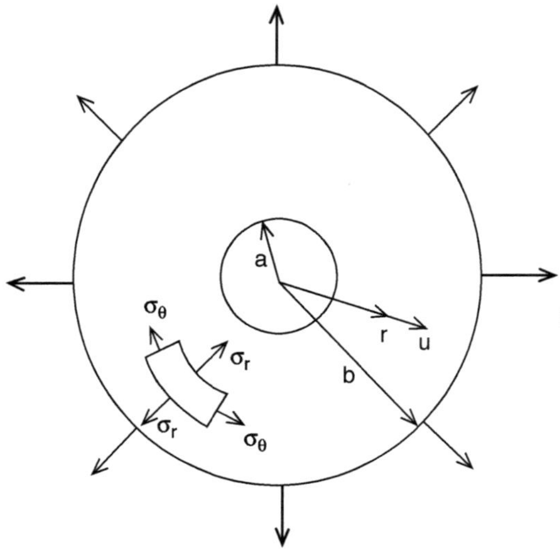

Fig. 5.23. Stretched plate

integrating it at h = constant as well as at boundary conditions $\sigma_r(a) = 0$, $\sigma_r(b) = p$ and excluding factor $(4u/3\Omega)^\mu$ we receive with the help of (5.129)

$$\sigma_\theta = p(1-\mu)(b/r)^\mu/(1-\beta^{\mu-1}) \tag{5.131}$$

where $\beta = b/a$. Putting σ_θ into (2.66) we find for the dangerous (internal) surface

$$e^{-\alpha\varepsilon}\varepsilon = 3\beta(1-\mu)^m\Omega(t)p^m/4(1-\beta^{\mu-1})^m \tag{5.132}$$

Applying to (5.132) criterion $d\varepsilon/dt \to \infty$ we have

$$\varepsilon_* = 1/\alpha, p^m\Omega(t_*) = 4(1-\beta^{\mu-1})^m/3\beta(1-\mu)^m\alpha e. \tag{5.133}$$

When the influence of time is negligible we compute from (5.133) at Ω = constant critical load

$$p_* = (4/3)^\mu(1-\beta^{\mu-1})/(1-\mu)(\alpha\beta\Omega e)^\mu. \tag{5.134}$$

At small μ that value must be compared to ultimate load p_u which follows from (5.131) at $\mu \to 0$ as

$$p_u = \sigma_{yi}(1-1/\beta)$$

where σ_{yi} is a yielding point at an axial tension or compression and the smallest value should be taken. At m near to unity we must compare p_* with yielding load which follows from (5.131) at m = 1 in form:

$$p_{yi} = (\sigma_{yi}/\beta)\ln\beta$$

and the consequent choice should be made. The solution may be also used for compression.

Sphere

For a sphere under internal q and external p pressures (Fig. 3.23) we denote the radial displacement also as u and according to relations (2.80) we compute $\varepsilon_\theta = u/\rho, \varepsilon_\rho = du/d\rho$ and from the constant volume demand (2.81) we find

$$u = C/\rho^2, \varepsilon_\rho = -2C/\rho^3, \varepsilon_\theta = C/\rho^3 \tag{5.135}$$

where constant C is to be established from boundary conditions. Now from (2.32) at $\alpha = 0$ and $\sigma_{eq} = \sigma_\theta - \sigma_\rho$ we deduce

$$\varepsilon_\theta = \Omega(t)(\sigma_\theta - \sigma_\rho)^m/2$$

or with consideration of (5.135)

$$\sigma_\theta - \sigma_\rho = (2C/\Omega\rho^3)^\mu. \tag{5.136}$$

Putting (5.136) into static equation (2.80) we get after integration at border demands $\sigma_\rho(b) = -p, \sigma_\rho(a) = -q$ and exclusion of constants

$$\sigma_\theta - \sigma_\rho = 3(q-p)\mu(b/\rho)^{3\mu}/2(\beta^{3\mu} - 1). \tag{5.137}$$

Now we use constitutive law (2.32) which for our structure is

$$e^{-\alpha\varepsilon}\varepsilon = 0.5\Omega(t)(\sigma_\theta - \sigma_\rho)^m \tag{5.138}$$

where $\varepsilon = \varepsilon_\theta$. Using here $\sigma_\theta - \sigma_\rho$ from (5.137) and criterion $d\varepsilon/dt \to \infty$ we deduce

$$\varepsilon_* = 1/\alpha, (q-p)^m\Omega(t_*) = 2(2m/3)^m(1-\beta^{-3\mu})^m/\alpha e. \tag{5.139}$$

When the influence of time is not high critical difference of the pressures at $\Omega =$ constant can be got

$$(q-p)_* = 2^{1+\mu}m(1-\beta^{-3\mu})/3(\alpha\Omega e)^\mu. \tag{5.140}$$

At small μ this value should be compared with $(q-p)_u$ according to (4.97) and the smaller one must be taken. Similar choice have to be fulfilled between $(q-p)_*$ and $(q-p)_{yi}$ given by (4.94) at μ near unity. At $p < 0$ the solution may be used for compression.

Cylinder

In an analogous way the fracture of a thick-walled tube can be studied. From (2.32) at $\alpha = 0, \varepsilon_x = 0, \varepsilon_\theta \equiv \varepsilon$ and $\sigma_{eq} = \sigma_\theta - \sigma_r$ we have

$$\varepsilon = (3/4)\Omega(t)(\sigma_\theta - \sigma_r)^m \tag{5.141}$$

and providing the procedure above for the disk and the sphere we find /17, 27/

$$\sigma_\theta - \sigma_r = 2\mu(q-p)(b/r)^{2\mu}/(\beta^{2\mu}-1). \quad (5.142)$$

Equation (2.32) for this structure is

$$e^{-\alpha\varepsilon}\varepsilon = 3\Omega(t)(\sigma_\theta - \sigma_r)^m/4. \quad (5.143)$$

Using here expression (5.142) at $r = a$ and the criterion $d\varepsilon/dt \to \infty$ we derive

$$\varepsilon_* = 1/\alpha, (q-p)^m\Omega(t_*) = 4(m/2)^m(1-\beta^{-2\mu})/3\alpha e. \quad (5.144)$$

When influence of time is negligible we can find as before critical difference of pressures as

$$(q-p)_* = (4/3)^\mu m(1-\beta^{-2\mu})/2(\alpha\Omega e)^\mu$$

and again for μ near to zero this value must be compared with $(q-p)_u$ according to (4.99) and smaller one have to be taken. The similar choice should be made between $(q-p)_*$ and $(q-p)_{yi}$ from relation (4.98) at m near unity.

From Fig. 5.24 where at $\alpha = 1, m = 1$ and $m = 2$ by solid and broken lines 1, 2, 3 for plate, sphere and cylinder curves $t_*(\beta)$ are represented respectively we can see that the critical time for a tube is less than the consequent one for the sphere and higher that of the plate.

Cone

We consider this task at the same suppositions as in Sects. 3.3.1 and 4.3.1. Using the scheme above, relations for strains (3.117) and stresses (3.116) as well as law (5.141) at $\sigma_r \to \sigma_\chi$ we find

$$\sigma_\theta - \sigma_\chi = (q-p)\cos^\mu\chi/J_6\sin^{2\mu}\chi \quad (5.145)$$

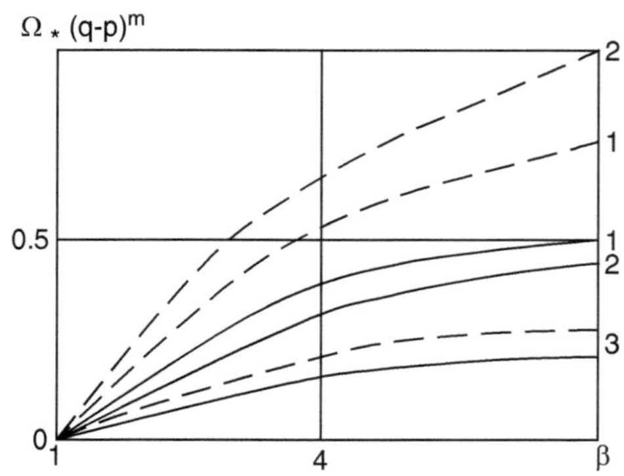

Fig. 5.24. Dependence of t_* on β and m

where

$$J_6 = \int_{\psi}^{\lambda} (\cos^{1+\mu}\chi / \sin^{2\mu+1}\chi) d\chi. \tag{5.146}$$

The computations for $m = 1$ when $J_6 = A/2$ from Sect. 3.3.1 at $\lambda = \pi/3$, $\psi = \pi/6$ show that integral J_6 can be easily calculated. For example at $m = 2$ and the shown meanings of λ, ψ its value is 0.9.

In order to appreciate the moment of fracture we put (5.145) into (5.143) and use criterion $d\varepsilon/dt \to \infty$ when we have for a dangerous (internal) surface

$$\varepsilon_* = 1/\alpha, \Omega(t_*)(q-p)^m = 4(J_6)^m(\sin^2\psi)/3e\alpha\cos\psi. \tag{5.147}$$

If the influence of time is negligible we derive from (5.147) at $\Omega =$ constant

$$(q-p)_* = (4/3)^{\mu} J_6(\sin^2\psi/\alpha e\Omega\cos\psi)^{\mu}. \tag{5.148}$$

Once more for small μ this value must be compared with $(q-p)_u$ according to (4.100) and smaller one should be taken. At μ near to unity $(q-p)_*$ must be compared to $(q-p)_{yi}$ from Sect. 4.3.1 and similar choice should be made.

Conclusion

The results of the solutions of this Section can be used for a prediction of a failure not only of similar structures but also of the voids of different form and dimension in soil massifs.

6

Ultimate State of Structures at Finite Strains

6.1 Use of Hoff's Method

6.1.1 Tension of Elements Under Hydrostatic Pressure

This approach takes as the moment of a fracture time t_∞ when the structures' dimensions become infinite. We consider as the first example a plate in tension by stresses p under hydrostatic pressure q (Fig. 6.1). Since here $\sigma_1 = \sigma_2 = p$, $\sigma_3 = -q$ we have from (2.31)

$$d\varepsilon/dt = 0.5B(p_o)^m(e^\varepsilon + \kappa_o)^m \tag{6.1}$$

where $\varepsilon = \varepsilon_1$, $\kappa_o = q/p_o$, and according to (1.42) $p = p_o e^\varepsilon$. The integration of (6.1) in limits $0 \leq \varepsilon \leq \infty$, $0 \leq 1 \leq t_\infty$ gives

$$B(p_o)^m t_\infty = 2\Bigg(\ln(1+\kappa_o) + (m-1)!\sum_{i=1}^{m-1}(m-1)!\sum(-1)^i!(1-1/(1+\kappa_o)^i)$$
$$/i!i(m-1-i)!\Bigg)/(\kappa_o)^m.$$

From Fig. 6.2 where for some κ_o curves $t_u(\mu)$ according to the latter expression are given by broken lines we can see that $t_u \equiv t_\infty$ diminishes with an increase of hydrostatic component.

In a similar way the fracture time can be found for a bar in tension by stresses q under hydrostatic pressure p (Fig. 6.3). In this case /17/ $\sigma_1 = q$, $\sigma_2 = \sigma_3 = -p$ and hence in (2.31)

$$S_1 = 2(p+q)/3, \ \sigma_{eq} = p + q.$$

Comparing this data to the previous ones we can see that rate $d\varepsilon/dt$ in the latter problem is twice of that for the plate. Hence t_∞ for the bar is one half of that in the plate case.

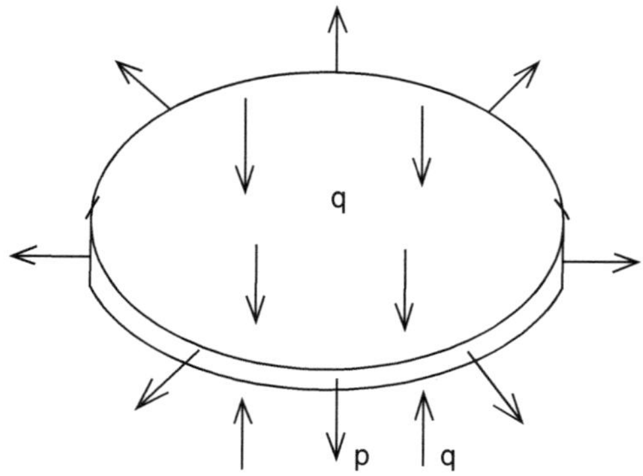

Fig. 6.1. Stretched plate under hydrostatic pressure

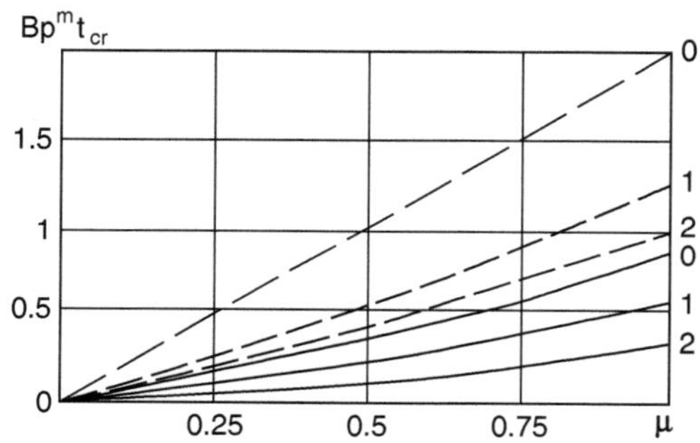

Fig. 6.2. Dependence of ultimate time t_{cr} on κ_o and μ

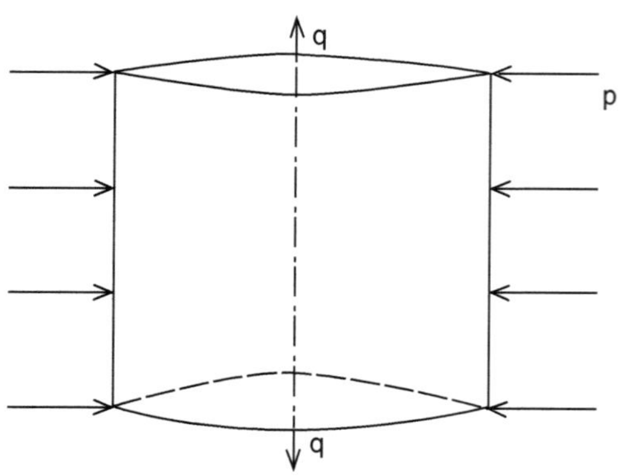

Fig. 6.3. Bar in tension under hydrostatic pressure

6.1.2 Fracture Time of Axisymmetrically Stretched Plate

In order to integrate differential equation (5.130) in the range of finite strains we take according to the condition of Tresca-Saint-Venant in Sect. 5.3.4. $r = r_o + u(t)$. We replace strains and displacements by their rates and rewrite (5.130) with consideration of (5.129) at B instead of Ω (see (2.31)) and $d\varepsilon/dt = dr/rdt$ as

$$d(r_o h_o \sigma_r)/dr_o = (4/3B)^{\mu} r_o h_o (r_o + u)^{-1-\mu}(du/dt)^{\mu}. \qquad (6.2)$$

Integration of (6.2) at boundary conditions $\sigma_r(a_o) = 0$, $\sigma_r(b_o) = p$ gives

$$h_b b_o p = (4/3B)^{\mu}(du/dt)^{\mu} \int_{a_o}^{b_o} r_o h_o(r_o)(r_o + u)^{-1-\mu} dr_o. \qquad (6.3)$$

Here $h_b = h_o(b_o)$. From (6.3) the fracture time can be found. Particularly at $h_o =$ constant and $h_o = h_b b_o/r_o$ we derive after transformations

$$B(b_o p)^m t_\infty = (4/3) \int_0^\infty (((b_o + u)^{1-\mu} - (a_o + u)^{1-\mu})/(1-\mu) + u((b_o + u)^{-\mu}$$
$$- (a_o + u)^{-\mu})/\mu)^m du,$$
$$Bp^m t_\infty = (4/3) m^m \int_0^\infty ((a_o + u)^{-\mu} - (b_o + u)^{-\mu})^m du. \qquad (6.4)$$

For any m integrals in (6.4) can be computed. If e.g. $m = 1$ we have respectively at $\beta = b_o/a_o$

$$Bpt_\infty = 4(1 - 1/\beta)/3, \quad Bpt_\infty = (4/3)\ln\beta \qquad (6.5)$$

and from Fig. 6.4 where broken lines 0, 1 are drawn according to (6.5) we can see that the curved profile has higher fracture time. Broken 2 and interrupted

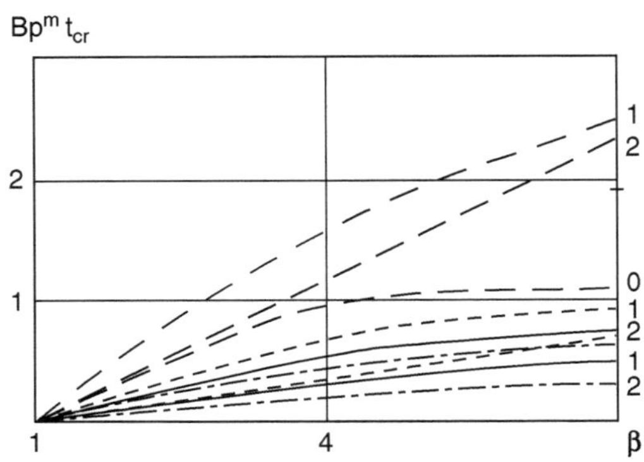

Fig. 6.4. Dependence of ultimate time t_{cr} on β and m for stretched plate

by points1 lines refer to the cases $m = 2$, $h_o = h_b b_o/r_o$ and h_o = constant when we derive from (6.4)

$$Bp^2 t_\infty = (16/3)\ln((1+\sqrt{\beta})^4/16\beta),$$
$$Bp^2 t_\infty = 16((3(1+\beta^2)/2 + 2\beta) - 2(1+\beta)\sqrt{\beta} + 2(\ln((1+\sqrt{\beta})/2) + \beta^2 \ln((1+1/\sqrt{\beta})/2))/3\beta^2.$$

6.1.3 Thick-Walled Elements Under Internal and External Pressures

We begin with a sphere and replace ε_θ, u in (5.135) by their rates $d\varepsilon_\theta/dt$, V. Then we suppose in (5.136) $\Omega(t) = Bt$. According to definition $\beta = b/a$ we have

$$d\beta/dt = (db/dt - \beta da/dt)/a \tag{6.6}$$

where

$$db/dt = V(b) = b d\varepsilon_\theta(b)/dt, \quad da/dt = V(a) = a d\varepsilon_\theta(a)/dt.$$

Using (5.136), (5.137) and (6.6) we derive after integration

$$B(q-p)^m t_\infty = 2(2m/3)^m \int_1^{\beta_o} ((\beta^{3\mu} - 1)^m/\beta(\beta^3 - 1))d\beta. \tag{6.7}$$

Here $\beta_o = b_o/a_o$ and critical time is equal to t_∞. Diagram $t_{cr}(\beta_o)$ according to (6.7) at $m = 1$ is given in Fig. 6.5 by broken line 0. The curves for $m = 2$, $m = 3$ go much higher. For them

$$B(q-p)^2 t_\infty = 128(\ln(\beta_o^{2\sqrt{3}} + 1/2\beta_o^{3/4}))/27,\ B(q-p)^3 t_\infty = 16(\ln\beta_o + 2\sqrt{3}\tan^{-1}((1-\beta_o)/\sqrt{3}(1+\beta_o))).$$

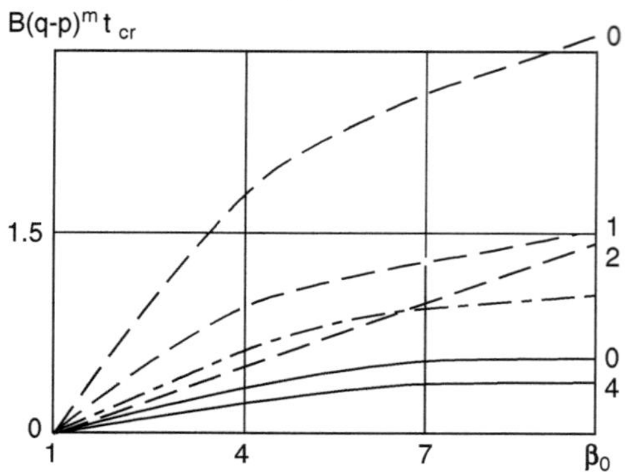

Fig. 6.5. Dependence of ultimate time t_{cr} on β_o and m for thick-walled elements

In a similar way the fracture of a cylinder can be considered. Using the procedure above for the disk and the sphere with a help of (6.6), (5.141), (5.142) we find

$$\sigma_\theta - \sigma_r = 2\mu(b/r)^2/(\beta^{2\mu} - 1), \tag{6.8}$$

$$B(q-p)t_\infty = (4/3)(m/2)^m \int_1^{\beta_o} ((\beta^{2\mu} - 1)^m/\beta(\beta^2 - 1))d\beta. \tag{6.9}$$

For $m = 1$ and $m = 2$ we compute respectively

$$B(q-p)t_\infty = (2/3)\ln\beta_o, \quad B(q-p)^2 t_\infty = (4/3)\ln((\beta_o + 1)^2/4\beta_o).$$

The consequent curves are drawn in Fig. 6.5 by broken lines 1, 2 and we can see that for the tube t_∞ is less than that for the sphere. Similarly for $m = 4$ we have (line goes higher)

$$B(q-p)^4 = (128/3)(\ln(\sqrt{\beta_o} + 1)^4/8\sqrt{\beta_o}(\beta_o + 1)).$$

For a cone we use the condition of its constant volume (4.101) for the whole body and introduce ratio

$$\beta = \cos\psi/\cos\lambda.$$

Then λ is function of β as

$$\cos\lambda = ((\beta_o - 1)/(\beta - 1))\cos\lambda_o$$

and integral J_6 in (5.146) is also a function of β. Now we find

$$d\beta/dt = (\beta - 1)(\tan\lambda)d\lambda/dt \tag{6.10}$$

and since (Fig. 3.24)

$$\varepsilon(\lambda) = \ln(\sin\lambda/\sin\lambda_o)$$

then

$$d\lambda/dt = \tan\lambda d\varepsilon(\lambda)/dt$$

and from (5.141) with $d\varepsilon/dt$, B instead of ε, Ω and (5.145) we derive

$$d\varepsilon(\lambda)/dt = (3B/4)(q-p)^m(\cos\lambda)/(J_6)^m \sin^2\lambda. \tag{6.11}$$

Putting $d\lambda/dt$ together with (6.11) into (6.10), separating the variables and integrating as before we have finally

$$B(q-p)^m t_\infty = (4/3)(\beta_o - 1)\cos\lambda_o \int_{\beta_o}^{1} (J_6)^m(\beta - 1)^{-2}d\beta. \tag{6.12}$$

The integrals in (6.12) should be calculated as a rule approximately.

6.1.4 Final Notes

Although the method in this sub-chapter uses somewhat unrealistic supposition of an infinite elongation at the rupture, sometimes the fracture time is near to test data. An analysis shows that the reason of it lays in the non-linearity of equations linking the rate of strains with stresses. Because of that the approach is widely used for the prediction of the failure moment of structures. For example in /17/ a row of elements are considered. Among them a grating of two bars, thin-walled sphere and tube under internal pressure, a long membrane loaded by hydrostatic pressure. Sometimes an initial plastic deformation is also taken into account. The task of axisymmetric thin-walled shells under the internal pressure is formulated. An attempt of consideration of stress change on the base of creep hypotheses is also made. But the method is mainly applied to a steady creep (see also Appendix L).

6.2 Mixed Fracture at Unsteady Creep

6.2.1 Tension Under Hydrostatic Pressure

For the bar in tension we use the notation of Fig. 6.3. According to relations (1.42), (1.45) we can link conditional q_o and true q stresses by expression

$$q = q_o e^{(1+\alpha)\varepsilon}$$

and from (2.32) we have

$$\varepsilon = \Omega(t)(q_o)^m (e^{(1+\alpha)\varepsilon} + k_o)^m. \tag{6.13}$$

By criterion $d\varepsilon/dt \to \infty$ we find from (6.13)

$$\varepsilon_* = \mu(\kappa_o \exp(-(1+\alpha)\varepsilon_*) + 1)/(1+\alpha).$$

We can get critical time t_* by putting ε_* into (6.13). As we can see from Fig. 6.6 the critical strains (at $\alpha = 0$) increase with a growth of the hydrostatic component.

In the same manner the failure of the plate in the axisymmetric tension under hydrostatic pressure (Fig. 6.1) can be studied. As a result we have in notation of Sect. 6.1.1

$$(p_o)^m \Omega(t_*) = 2\varepsilon_* (\exp(1+\alpha)\varepsilon_* + \kappa_o)^{-m}$$

and as we can see from Fig. 6.2 where for a steady creep ($\Omega = Bt$) by solid lines at the same κ_o as the Hoff's method diagrams $t_*(\mu)$ are constructed critical time increases (similar to fracture time) with a fall of the hydrostatic component. We can also notice that $t_* < t_\infty$ and it can be shown (see Appendix L) that with a growth of creep curves non-linearity the difference between critical and fracture times increases. It can be explained first of all by the circumstance that the method of infinite strain rate takes more realistic condition of failure at finite strains than the Hoff's approach.

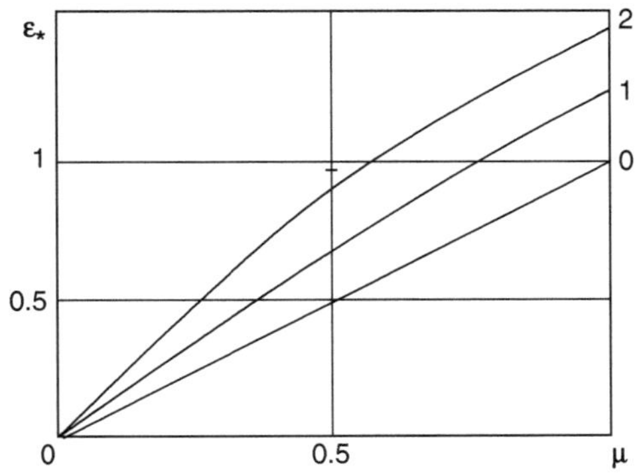

Fig. 6.6. Dependence of ε_* on μ at different κ_o for bar in tension under hydrostatic pressure

6.2.2 Axisymmetric Tension of Variable Thickness Plate with Hole

General Case and that of Constant Thickness

Here /29/ we use the same suppositions as in Sect. 6.1.2 which bring (2.32) to form

$$\varepsilon e^{-\alpha\varepsilon} = (3/4)\Omega(t)(\sigma_\theta)^m$$

where $\varepsilon = \varepsilon_\theta = \ln(r/r_o)$. Using the condition of a constant volume as

$$(r_o + u)h = r_o h_o$$

with $u = u(t)$ we integrate differential equation (5.130) in initial variables as follows:

$$(3\Omega/4)^\mu h_b b_o p = \int_{a_o}^{b_o} h_o(r_o)(1 + u/r_o)^{-1-\alpha\mu} \ln^\mu(1 + u/r_o) dr_o. \qquad (6.14)$$

If we seek the critical state with the help of a computer we can apply criterion $d\varepsilon/dt \to \infty$ directly to expression (6.14). Calculated in this way diagrams $\varepsilon_{b*}(\beta)$ (here $\beta = b_o/a_o$), h_o = constant, $\alpha = 0$ and $\alpha = 1$ are represented in Fig. 6.7 by solid and broken lines 1. Critical time t_* for these cases at $\Omega = Bt$ is given from (6.14) in Fig. 6.4 by solid curves 2 and 1.

Curved Profile

For the case $h_o = h_b b_o/r_o$, $\alpha = 0$ we derive from (6.14) at $du/dt \to \infty$ equality

$$\int_{a_o}^{b_o} (1 + u/r_o)^{-2}(\mu \ln^{\mu-1}(1 + u/r_o) - \ln^\mu(1 + u/r_o))(r_o)^{-2} dr_o = 0$$

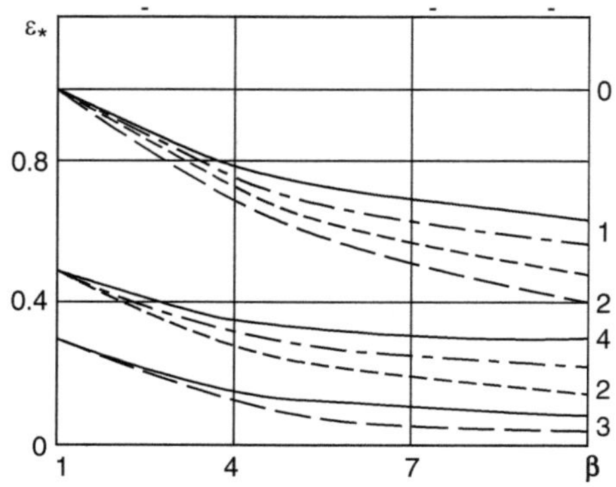

Fig. 6.7. Dependence of ε_* on β for thick-walled structures

or after computing the integral

$$(1 + \xi_*/\beta)\ln^{\mu}(1 + \xi_*) = (1 + \xi_*)\ln^{\mu}(1 + \xi_*/\beta)$$

where $\xi = u/a_o$. This equation can be solved parametrically if we suppose

$$1 + \xi_*/\beta = (1 + \xi_*)^{\eta}.$$

Here η is a parameter. We have from this equation

$$\xi_* = \eta^{\mu/(\eta-1)} - 1$$

and we can find other variables in form

$$\beta = \xi_*/((1 + \xi_*)^{\eta} - 1), \quad \varepsilon_{b*} = \eta\ln(1 + \xi_*).$$

Curves $\varepsilon_{b*}(\beta)$ are given by dotted lines 1 and 2 in Fig. 6.7 for m = 1 and m = 2 respectively. When ξ_* is known we can find t_* from (6.14) as

$$p^m\Omega(t_*) = (4/3)\left(\int_1^{\beta}(\xi_* + \rho)^{-1}\ln^{\mu}(1 + \xi_*/\rho)d\rho\right)^m$$

where $\rho = r_o/a_o$. Diagrams $t_*(\beta)$ are drawn by dotted lines 1 and 2 in Fig. 6.4 for m = 1 and m = 2 respectively. We can see that in this case the curved profile also gives higher critical time than that with h_o = constant. This indicates that an optimal profile can be searched.

Optimal Profile

We shall seek such a disk among ones with radial cross-sections as following

$$h_o = a_o(\beta - 1)^{-1}((h_a - \beta h_b)b_o/(r_o)^2 + (\beta^2 h_b - h_a)/r_o). \tag{6.15}$$

Putting this expression into (6.14) at $\alpha = 0$ we receive

$$(3\Omega/4)^{\mu} h_b \beta p = ((\beta^2 h_b - h_a) \int_{a_o}^{b_o} (1 + u/r_o)^{-1} \ln^{\mu}(1 + u/r_o) dr_o/r_o + (b_o(h_a - \beta h_b) + (\ln^{1+\mu}(1 + u/a_o) - \ln^{1+\mu}(1 + u/b_o))/u(1 + \mu))/(\beta - 1). \quad (6.16)$$

Using criterion $du/dt \to \infty$ and integrating we find equation

$$(1 + \mu)(1 - 1/\beta)\xi_*(h_a(1 + \xi_*)^{-1} \ln^{\mu}(1 + \xi_*) - \beta h_b(1 + \xi_*/\beta)^{-1} \ln^{\mu}(1 + \xi_*/\beta)) = (h_a - \beta h_b)(\ln^{1+\mu}(1 + \xi_*/\beta) - \ln^{1+\mu}(1 + \xi_*)).$$

from which ratio h_a/h_b can be computed and we consider limit case $h_a = 0$. For it we have at $\Omega = Bt$

$$\ln^{1+\mu}(1 + \xi_*) - \ln^{1+\mu}(1 + \xi_*/\beta) + (1 + \mu)(1 - !/\beta)\xi_*(1 + \xi_*/\beta)^{-1} \times \ln^{\mu}(\xi_*/\beta) = 0,$$

$$Bp^m t_* = 4 \left(\int_1^{\beta} (\rho + \xi_*)^{-1} \ln^{\mu}(1 + \xi_*/\rho) d\rho + (\ln^{1+\mu}(1 + \xi_*/\beta) - \ln^{1+\mu}(1 + \xi_*))/(1 + \mu)\xi_* \right)^m /3\,(1 - 1/\beta)^m .$$

Diagrams $\varepsilon_{b*}(\beta)$ for $m = 1$ and $m = 2$ are drawn in Fig. 6.7 as interrupted by points lines 1 and 2. We can see that they are somewhat higher than the consequent values for profile $h_o = h_b b_o/r_o$. The curves $t_*(\beta)$ for $m = 1$ and $m = 2$ are shown also by the same lines 1 and 2 as in Fig. 6.4 and they are lower than respective ones for curved profile above. However all the curves for every m are near to each other and from practical point of view the disk h_o = constant should be recommended (the curves $h_b = 0$ are the lowest). Similar solution may be received for rotating disk /30/ (see Appendix O).

6.2.3 Thick-Walled Elements Under Internal and External Pressures

Sphere

For this structure the initial equations are

$$\varepsilon_{\chi} = \varepsilon_{\theta} \equiv \varepsilon = \ln(\rho/\rho_o) \equiv (1/3)\ln(\rho/\rho_o)^3,$$
$$\rho^3 - a^3 = (\rho_o)^3 - (a_o)^3, \quad \varepsilon e^{-\alpha\varepsilon} = \Omega(t)(\sigma_{\theta} - \sigma_{\rho})^m/2. \quad (6.17)$$

Here ρ, $\chi = \theta$ are spherical coordinates. The second of expressions (6.17) is the condition of constant volume and using it for the whole sphere ($\rho = b$)

as well as the first expression (6.17) we represent the strains at internal and external surfaces as

$$\varepsilon_a = (1/3)\ln\kappa, \quad \varepsilon_b = (1/3)\ln\kappa_b$$

where

$$\kappa = (a/a_o)^3, \quad \kappa_b - 1 = (\kappa - 1)/(\beta_o)^3, \beta_o = b_o/a_o.$$

In order to integrate the static equation (2.80) in current variables we write the second (6.17) as following:

$$1 - (\rho_o/\rho)^3 = 1 - e^{-3\varepsilon} = (a^3 - (a_o)^3)/\rho^3$$

and differentiate it. That gives

$$d\rho/\rho = (1 - \exp 3\varepsilon)^{-1} d\varepsilon.$$

Putting the last expression together with the third (6.17) into the first (2.80) we determine after integration

$$(q - p)\Omega(t) = 2^{1+\mu} \int_{\varepsilon_b(\kappa)}^{\varepsilon_a(\kappa)} (\exp 3\varepsilon - 1)^{-1} (e^{-\alpha\varepsilon}\varepsilon)^{\mu} d\varepsilon. \tag{6.18}$$

According to criterion $d\varepsilon/dt \to \infty$ we receive from (6.18)

$$\kappa_*^{-(1+\alpha\mu/3)} \ln^{\mu}\kappa_* = \kappa_{b^*}^{-(1+\alpha\mu/3)} \ln^{\mu}\kappa_{b^*}$$

and if we suppose $\kappa_* = \kappa_{b*}^{\eta}$ the results can be represented as functions of parameter η

$$\kappa_{b^*} = \eta^{\mu/(\eta-1)(1+\alpha\mu/3)},$$

after that we find κ_* and

$$\beta_o = ((\kappa_* - 1)/(\kappa_{b^*} - 1))^{1/3}.$$

Considering in (6.18) the critical values of strains we get on the critical time. From Fig. 6.7 where for $m = 1$, $\alpha = 0$ and $\alpha = 0.9$ diagrams $\varepsilon_{b^*}(\beta_o)$ are represented by solid and broken lines 3 respectively we can see that the critical strains decrease with a growth of α and β_o. In certain conditions (big α, β_o, m) the strains at fracture can be small enough and it explains brittle destruction of structures made of plastic materials.

Comparing solid curve 3 in Fig. 6.7 with straight line 0 that corresponds to simple (see (1.46) at $\alpha = 0$, $m = 1$) or biaxial equal tension we can see that even thin-walled (β_o is near to unity) sphere failures at strains three times less than a plate in the same stress state. From Fig. 6.5 where function $t_*(\beta_o)$ is shown by solid line 0 for the case of sphere at $\alpha = 0$, $m = 1$ we can see that t_* is less than t_∞.

Cylinder and Cone

In the same manner we consider the failure of a thick-walled tube /30/. The results can be represented in a form similar to (6.18) as

$$(q-p)^m\Omega(t_*) = (4/3)\left(\int\limits_{\varepsilon(\kappa_{b*})}^{\varepsilon(\kappa_*)} (e^{2\varepsilon}-1)^{-1}(\varepsilon e^{-\alpha\varepsilon})^{\mu} d\varepsilon\right)^m \tag{6.19}$$

wherein $\varepsilon = \varepsilon_\theta$, $\kappa = (a/a_o)^2$, $\kappa_b = 1+(\kappa-1)/(\beta_o)^2$ and according to criterion $d\varepsilon/dt \to \infty$ we have

$$\kappa_{b*}\ln^{\mu}\kappa_* = \kappa_*\ln^{\mu}\kappa_{b*}.$$

A solution of the latter equation with a help of parameter η as in the cases of the disk and the sphere is also possible. Diagrams $\varepsilon_*(\beta)$ and $t_*(\beta)$ are drawn in Figs. 6.7 and 6.5 by solid lines 4.

A comparison of the results for the sphere and the cylinder allows to conclude that the fracture of the first one demands longer time than that of the second. However the strains at external surface in unstable state of the sphere are less than those of the cylinder and the latter are smaller for the plate in biaxial equal tension. All that can be explained by the influence of the form of a structure that is not taken into account by classical approaches to finding the ultimate state by the strength hypotheses.

In order to use the criterion of infinite strains rate to a cone we write the condition of the constant material's volume (4.101) in form

$$\cos\chi - \cos\chi_o = (\delta-1)\cos\psi_o$$

where $\delta = \cos\psi/\cos\psi_o$. From these expressions and the definition of tangential strain (Fig. 3.24) as

$$\varepsilon(\chi) = \ln(\sin\chi/\sin\chi_o)$$

we derive equation

$$(1-e^{-2\varepsilon})\sin^2\chi + 2(\delta-1)\cos\psi_o\cos\chi - (\delta-1)^2\cos^2\psi_o = 0$$

that allows to determine $\cos\chi$ by solution

$$\cos\chi = (1-e^{-2\varepsilon})^{-1}(\delta-1)\cos\psi_o + \sqrt{1+(1-e^{-2\varepsilon})^{-2}e^{-2\varepsilon}(\delta-1)^2\cos^2\psi}$$

and $d\chi$ is proportional to $d\varepsilon$. With consideration of the last dependence and the constitutive law we integrate (3.116) as follows:

$$(3\Omega/4)^{\mu}(q-p) = \int\limits_{\varepsilon_b(\delta)}^{\varepsilon_a(\delta)} \varepsilon^{\mu-1}e^{-(2+\alpha\mu)}((1-\cos\chi)^{-1}\cos\chi d\varepsilon).$$

Applying to this relation criterion $d\varepsilon/dt \to \infty$ we get on an equation for critical value of δ consideration of which gives the critical time. If its influence is negligible we have a critical difference of pressures as before /31/.

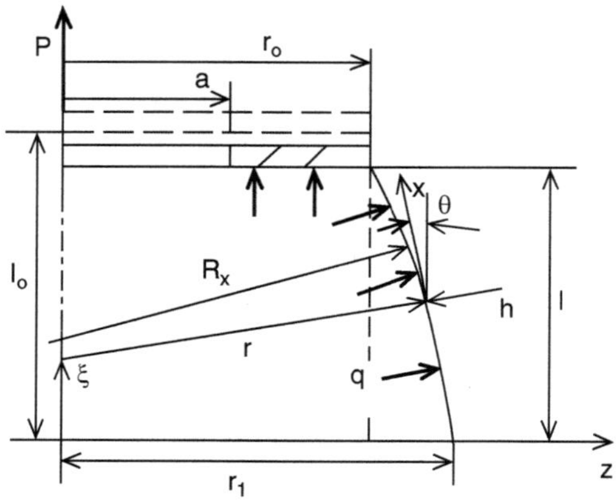

Fig. 6.8. Thin-walled shell

6.2.4 Deformation and Fracture of Thin-Walled Shells Under Internal Pressure

General Relations

In the following we are going to consider a shell /32/ with initial length $2l_o$ between rigid bottoms with coaxial holes of radius a (in Fig. 6.8 a quarter of the structure is shown), diameter $2r_o$, the wall thickness h_o under axial force P and internal pressure q. We write down basic expressions for the shell's deformed state (solid lines in the figure). Static laws are:

$$\sigma_x = (q(r^2 - a^2) + P/\pi)/2rh\cos\theta, \quad d(hr\sigma_x)/dr = h\sigma_y \tag{6.20}$$

where σ_x, σ_y are stresses in longitudinal and circumferential directions, θ – angle between the shell's ξ and x axes; relations for finite strains

$$\varepsilon_x = \ln(dr/\sin\theta d\xi), \ \varepsilon_y = \ln(r/r_o), \ \varepsilon_z = \ln(h/h_o) \tag{6.21}$$

which are linked by constant volume condition

$$\varepsilon_x + \varepsilon_y + \varepsilon_z = 0.$$

The system is closed by a rheological expression which we take with a consideration of the stress state at $P = 0$ as $\varepsilon_x = 0$. We suppose the second constitutive equation in form like (1.45) and (2.30)

$$2\tau_e = \omega(t)(\gamma_m)^\mu \exp(-\alpha\varepsilon_1). \tag{6.22}$$

Equations (6.21) can be replaced by expression

$$v_o = \int_0^{\varepsilon_m} f(\varepsilon_1)(\sin\theta)^{-1}d\varepsilon_1. \tag{6.23}$$

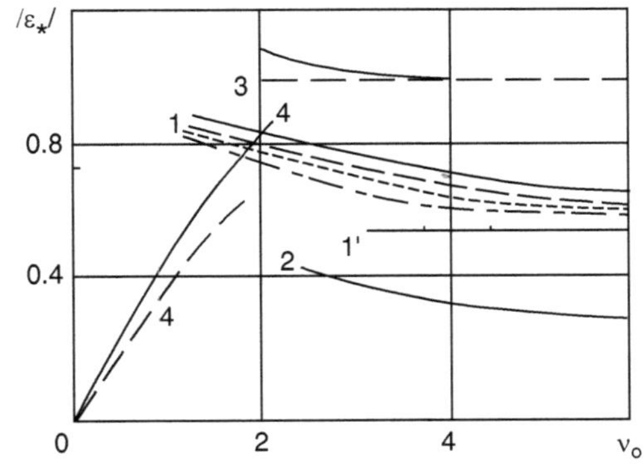

Fig. 6.9. Dependence of critical strain on relative length of shell

Here $f = \exp \varepsilon_1$, $\nu_o = l_o/r_o$, $\varepsilon_1 = \varepsilon_y = \varepsilon$, $\varepsilon_m = \varepsilon$ at $\xi = 0$ and function $\theta(\varepsilon_m, \varepsilon)$ can be derived from (6.20), (6.21) in form

$$(e^{2\varepsilon} - \chi_o)/\cos\theta = \exp 2\varepsilon_m - \chi_o + N_q \int_{\varepsilon_m}^{\varepsilon} \varepsilon^{\mu} e^{-\alpha\varepsilon} d\varepsilon \qquad (6.24)$$

where $\chi_o = (a/r_o)^2$, $N_q = 2^{1+\mu}\omega(t)h_o/qr_o$.

With consideration of (6.24) we apply criterion $d\varepsilon_m/dt \to \infty$ to (6.23). The consequent diagram $\varepsilon_*(\nu_o)$ for $m = 1$, $\alpha = \chi_o = 0$ is given by solid line 1 in Fig. 6.9. Computations show that at $\chi_o = 1$ (the case of rings) the consequent curve is near to that line. Critical time t_* can be found from expression for N_q at $\varepsilon = \varepsilon_*$.

Some Approximate Solutions

We demonstrate one of them for the case $\chi_o = 0$ when from (6.24) we have at $\varepsilon_m - \varepsilon = \eta$ (η is a small value) after decomposition of exponents in series and neglecting members with η in power 2 and more

$$1/\cos\theta = 1 + \eta(2 - N_q \varepsilon_m^{\mu} \exp(-(2+\alpha)\varepsilon_m)),$$

and from (6.23)

$$\nu_o = \sqrt{2\varepsilon_m}(\exp \varepsilon_m)/\sqrt{2 - N_q \varepsilon_m^{\mu} \exp(-(2+\alpha)\varepsilon_m)}.$$

Using the criterion $d\varepsilon_m/dt \to \infty$ we compute

$$N_{q*} = 2(1 + 2\varepsilon*) \exp((2 + \acute{\alpha})\varepsilon_*)/\varepsilon_*^{\mu}(1 - \mu + (4 + \acute{\alpha})\varepsilon_*)^{0.5},$$
$$\nu_o = \sqrt{\varepsilon_*}(\exp \varepsilon_*)\sqrt{1 - \mu + (4+\alpha)\varepsilon_*}/\sqrt{(2+\alpha)\varepsilon_* - \mu}.$$

For case $\mu = 1$, $\alpha = 0$ we have simple expressions

$$N_{q*} = (1 + 2\varepsilon_*)(\exp 2\varepsilon_*)/\varepsilon_*^2,$$
$$\nu_o = 2\varepsilon_*(\exp \varepsilon_*)/\sqrt{2\varepsilon_* - 1}.$$

The curve $\nu_o(\varepsilon_*)$ is inside the totality of lines 1 When $\nu_o \to \infty$ (infinite length) we derive $\varepsilon_* = 1/2m$. If $\mu = 0.5$ the consequent diagram is constructed in Fig. 6.9 by solid line 2 and it is below the respected curve for $m = 1$.

Another Approximate Approach

In some works (/32, 33/ and others) the hypothesis is taken that straight lines in x directions deform as circular arcs with radius R_x (Fig. 6.8) defined by expression

$$R_x = 0.5((r_1 - r_o)^2 + l^2)/(r_1 - r_o).$$

That supposition fully determines the geometry and the deformation of the shell. Particularly

$$\cos\theta = (R_x - r_1 + r)/R_x,$$
$$\varepsilon_1 = \ln(1 + 2\rho \sin^2(\nu_o/2\rho))$$

where $\rho = R_x/r_o$. Computing from (6.20), (6.21) σ_y and putting it together with ε_1 into (6.22) we find expression for ρ as follows:

$$N_q 2\rho \ln^{\mu}(1 + 2\rho \sin^2(\nu_o/2\rho)) = (2\rho\cos^2(\nu_o/2\rho) - 1)(1 + 2\rho\sin^2(\nu_o/2\rho))^{2+\alpha}.$$

Using criterion $d\rho/dt \to \infty$ and excluding N_q we derive an equation for ρ_*

$$(\nu_o \sin(\nu_o/\rho_*) + 1)\exp\varepsilon_* = 2\rho_*^2(\mu/\varepsilon_* - 2 - \alpha)(2\cos^2(\nu_o/2\rho_*) - 1/\rho_*)$$
$$\times (\sin^2(\nu_o/2\rho_*) - (\nu_o/\rho_*)\sin(\nu_o/\rho_*))$$

according to which interrupted by points line1 is given in Fig. 6.9. The dotted curve is constructed in the supposition of $l = l_o$ /33/. The good agreement of these results with the rigorous solution (especially at $\nu > 5$) opens the way for approximate study of other shells.

Torus of Revolution

We suppose that the structure (one quarter of which is shown in Fig. 6.10) changes at deformation its dimensions but not the form. The measurements and natural observations show that

$$R_o - r_o = R - r$$

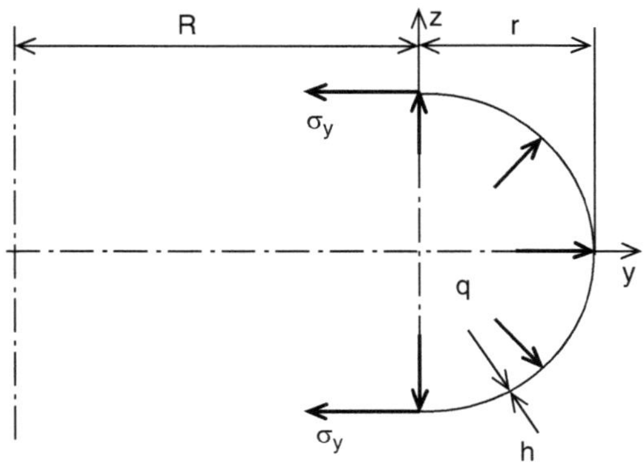

Fig. 6.10. Torus of revolution

and the fracture takes place in points y = 0, z = r. So, for them at h = constant we have

$$Rhr = R_o h_o r_o$$

(this expression is valid also for the whole volume of the material at h_o = constant). The stresses there are

$$\sigma_y = \sigma_o e^{2\varepsilon}(\rho_o + e^{\varepsilon})/(\rho_o + 1)$$

where $\sigma_o = qr_o/h_o$, $\rho_o = R_o/r_o - 1$, $0 < \rho_o < \infty$ and strain $\varepsilon = \varepsilon_1 = \varepsilon_y$ is determined by the second expression (6.21).

Deformation ε_z is given by the third relation under this number and for ε_x we write

$$\varepsilon_x = \ln(R/R_o),$$

So, with consideration of constant volume condition we find maximum shear

$$\gamma_m = 2\varepsilon + \ln((\rho_o + e^{\varepsilon})/(\rho_o + 1))$$

and rheological law (6.22) becomes

$$\sigma_o e^{(2+\alpha)\varepsilon}(\rho_o + e^{\varepsilon}) = \omega(t)(2\varepsilon + \ln((\rho_o + e^{\varepsilon})/(\rho_o + 1)))^{\mu}(\rho_o + 1).$$

According to criterion $d\varepsilon/dt \to \infty$ we derive equation for critical deformation as

$$((2+\alpha)(\rho_o + \exp\varepsilon_*) + \exp\varepsilon_*)(2\varepsilon_* + \ln(\rho_o + \exp\varepsilon_*))/(\rho_o + 1) = \mu(2\rho_o + 3\exp\varepsilon_*).$$

At $\rho_o \to \infty$ (a long tube) and $\rho_o = 0$ (a sphere) with replacement of α by $\mu\alpha$ we compute respectively the relations that were received earlier for these structures

$$\varepsilon_* = 1/(2m + \alpha), \quad \varepsilon_* = 1/(3m + \alpha)$$

and that confirms a validity of the hypotheses above. The consequent curves for $\alpha = 0$ are drawn in Fig. 6.11 by solid straight lines t, s and we can expect that other cases are between them.

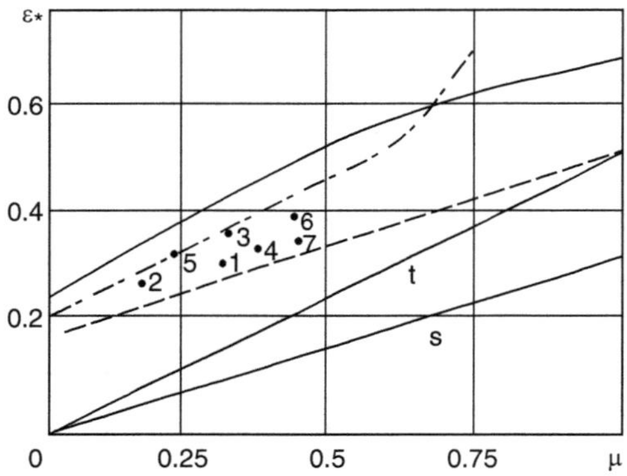

Fig. 6.11. Dependence of critical deformations on μ

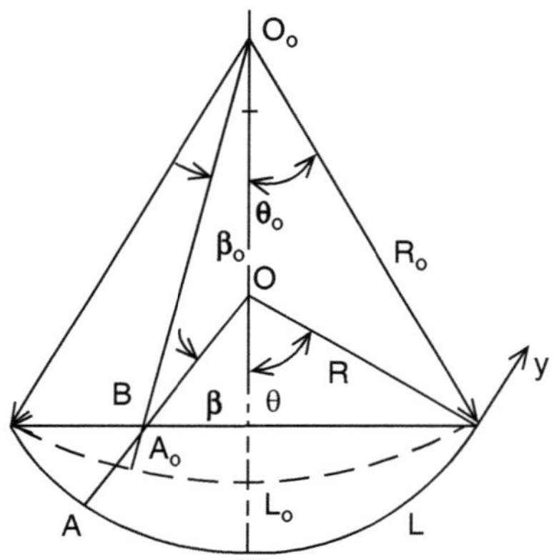

Fig. 6.12. Cross-section of membrane

6.2.5 Thin-Walled Membranes Under Hydrostatic Pressure

General Expressions and Cylindrical Membrane

These structures are more often used in the Geo-mechanics (see e.g. /34/, Appendix N). We begin with a long membrane as a part of a cylinder the cross-section of which is shown in Fig. 6.12 and suppose as before that its form remains at a deformation (solid line in the figure). From the condition of the constant material's volume and geometrical considerations we have

$$Lh = L_o h_o,\ L_o = R_o \theta_o, L = R\theta,\ R_o \sin\theta_o = R \sin\theta \qquad (6.25)$$

where h is a thickness of the membrane. We shall solve the problem in terms of angle θ when for $\varepsilon \equiv \varepsilon_y$ and $\sigma_y = 2\tau_e$ we receive with consideration of (6.25)

$$\varepsilon = \ln(\theta \sin\theta_o / \theta_o \sin\theta), \sigma_y = \sigma_o \theta \sin^2\theta_o / \theta_o \sin^2\theta.$$

Here $\sigma_o = qR_o/h_o$ and q is the hydrostatic pressure. Putting ε and σ_y into (6.22) we deduce

$$2\theta \sin^2\theta_o/\theta_o \sin^2\theta = N_q \ln^\mu(\theta \sin\theta_o/\theta_o \sin\theta)$$

where N_q has the same value as in (6.24) with replacement r by R_o.

Using criterion $d\theta/dt \to \infty$ and excluding N_q we find equation for critical θ

$$\mu = (\sin\theta_* - 2\theta_* \cos\theta_*)/(\sin\theta_* - \theta_* \cos\theta_* + \alpha)\ln(\theta_* \sin\theta_o/\theta_o \sin\theta_*)$$

which for $\theta_o = \theta_* = \pi$ (a cylinder) gives at $\alpha = 0$, $\varepsilon_* = \mu/2$ as for a long tube (solid straight line t in Fig. 6.11) and for another ultimate case $\theta_o = 0$ we have the solid curve. So we can conclude that at $0 < \theta < \pi$ the points are between these lines and the influence of the form of the structure on its ultimate state is high.

Spherical Membrane

For such a segment we keep picture in Fig. 6.12 and take condition

$$\varepsilon \equiv \varepsilon_x = \varepsilon_y = -\varepsilon_z/2.$$

Then with consideration of (6.25) we write for stresses

$$\sigma \equiv \sigma_y = \sigma_x = qR_o e^{2\varepsilon}(\sin\theta_o/\sin\theta)/2h_o.$$

Putting σ and ε into (6.22) we derive

$$qR_o(\sin\theta_o/\sin\theta)/h_o = 3^\mu 2\varepsilon^\mu e^{-(2+\alpha)\varepsilon}\omega(t).$$

According to criterion $d\varepsilon/dt \to \infty$ and after an exclusion of $\omega(t)$ together with the constants we find

$$\mu = (2 + \alpha - (d\theta/d\varepsilon)_* \cot\theta_*)\varepsilon_*. \qquad (6.26)$$

Further we can take different suppositions. The simplest of them is h = constant, but strains can be calculated in two options. From the condition of material incompressibility we have

$$\varepsilon = 0.5\ln((1 + \cos\theta_o)/(1 + \cos\theta))$$

and from (6.26)

$$\mu = (0.5\alpha + (1 - 2\cos\theta_*)/(1 - \cos\theta_*))\ln((1 + \cos\theta_o)/(1 + \cos\theta_*)).$$

If we compute ε according to a change of a meridian then similar to the case of the cylindrical membrane we receive

$$\mu = (\alpha + (2\tan\theta_* - 3\theta_*)/(\tan\theta_* - \theta_*))\ln(\theta_* \sin\theta_o/\theta_o \sin\theta_*). \qquad (6.27)$$

The calculations show that both options give similar results that for the case $\alpha = \theta_o = 0$ are constructed by broken curve in Fig. 6.11. Its comparison to straight line s for the sphere shows also the big influence of the form on the critical state of this structure.

The case of non-homogeneous deformation (when h is not a constant) we consider on the base of the supposition that point A_o on straight line O_oB in Fig. 6.12 comes to point A of ray OB. From the figure we find the expression (valid at $\theta_o < \pi/2$) for ε in the pole where $\beta = \beta_o = 0$

$$\varepsilon = -\ln(\cos\theta/\cos\theta_o)$$

putting which into (6.26) we receive

$$\mu = -(2 + \alpha - \cot^2\theta_*)\ln(\cos\theta_*/\cos\theta_o). \tag{6.28}$$

The consequent curve is drawn in Fig. 6.11 by interrupted by points line at $\alpha = 0$.

Comparison with Test Data

In Fig. 6.11 the experimental points are given according to /35/ for mild copper (1 – Brown-Sacks), aluminum, hard and mild steel (2, 3, 4 – Brown-Tomson), two types of copper (5, 6 – Weil-Newmark), polyethylene (7 from /36/). We can see that these points are situated between curves according to (6.28), (6.27). As the first of them gives safer values it can be recommended for practical use. We can also note that solution (6.27) at $\mu = 1$, $\alpha = \theta_o = 0$ gives θ_* near to represented in /37/ where deformation $\varepsilon = \theta/\sin\theta$ and linear link between σ, ε are taken. This θ_* corresponds well enough to test data on butadiene rubber membranes of different thickness. It opens the way to the theory above also for rubbers.

So, the test data above confirm the theory in the case of a spherical segment. The good agreement of this approach was fulfilled by the author for thin-walled tubes under axial load, internal pressure and torsion (see /14, 38/ and others).

6.2.6 Two other Problems

Tension of Limited Length Tube

Now we consider the case of axial load P action when we suppose that in dangerous cross- section $\xi = 0$ the strains are linked as

$$\varepsilon_y = \varepsilon_z = -\varepsilon_x/2$$

(a hypothesis of the material). We rewrite the second equilibrium expression (6.20) with consideration of (6.21) as follows

$$2\exp\varepsilon_x d(\exp(-\varepsilon_x)\sigma_x)/d\varepsilon_x = -\sigma_y$$

or taking into account (6.22) with $2\tau_e = \sigma_x - \sigma_y$

$$d(\sigma_x \exp(-3\varepsilon_x/2)) = 0.5(3/2)^\mu \omega(t)\exp(-(\alpha+0.5)\varepsilon_x)d\varepsilon_x.$$

An integration of this expression together with (6.21) and the first (6.20) gives at $\varepsilon_x \equiv \varepsilon$

$$1/\cos\theta = e^{-\varepsilon/2}\left(\exp(\varepsilon_m/2) + N_P\int_{\varepsilon_m}^{\varepsilon}\varepsilon^\mu e^{-(\alpha+0.5)\varepsilon}d\varepsilon\right) \tag{6.29}$$

where

$$N_P = (3/2)^\mu\omega(t)/\sigma_o, \quad \sigma_o = P/2\pi r_o h_o.$$

Now we apply criterion $d\varepsilon/dt \to \infty$ to (6.23) with $f(\varepsilon_1) = 0.5\exp(-3\varepsilon/2)$. Computed with a help of a PC curve for $\alpha = 0$, $m = 1$ is given in Fig. 6.9 by solid line 3 in which broken straight line corresponds to the bar of infinite length. We can see that the influence of the latter takes place only for small v_o. Critical time t_* can be found from the expression for N_P at $\varepsilon_m = \varepsilon_*$.

Compression of Cylinder

The solution above can be used for a compression by rough plates of a short cylinder (see Fig. 4.38) if we suppose that it consists of a set of thin-walled tubes. Computed with a help of PC diagram is represented by solid line 4 in Fig. 6.9. The difficulties of this task solution and its importance allow to apply some approximate approaches similar to that for the thin-walled shell under internal pressure above.

We rewrite approximately (6.29) as

$$1/\cos\theta = 1 + 0.5\eta(1 - N_P\varepsilon_m^\mu(\exp(-(1+\alpha)\varepsilon_m))).$$

Putting it in (6.23) we receive after integration

$$v_o = \sqrt{\varepsilon_m}\exp(-1.5\varepsilon_m)/\sqrt{1 - N_p(\varepsilon_m^\mu\exp(-(1+\alpha)\varepsilon_m))}.$$

Using criterion $d\varepsilon/dt \to \infty$ we finally find

$$N_{P*} = (3\varepsilon_* - 1)\exp((1+\alpha)\varepsilon_*)/\varepsilon_*/^{-\mu}/(/\mu - 1 + (2-\alpha)\varepsilon_*/),$$
$$v_o = \exp(-1.5\varepsilon_*)\sqrt{/\mu - 1 + (2-\alpha)\varepsilon_*//\varepsilon_*//}\sqrt{/\mu - (1+\alpha)\varepsilon_*./}$$

At $\alpha = 0$, $\mu = 1$ the curve in tension practically coincides with line 3 in Fig. 6.9 and at compression ($\varepsilon_* < 0$) broken curve 4 is near to that of the rigorous solution above., It is interesting to notice that the approximate results are valid for big ε_m. The critical time may be found from expression for N_{P*}.

Final Notes

The relations above are important because the theory supposes infinite length of samples although the experiments are usually made on the ones of limited length and so we can appreciate its influence on the results.

6.2.7 Ultimate State of Anisotropic Plate in Biaxial Tension

General Considerations

The application of this sub-chapter method to a plate in a tension allows not only to find the ultimate state of similar structure elements but also to establish a theoretical strength of an element of a material. The latter problem is usually solved in an experimental way with a help of the so-called strength hypotheses.

Basic Expressions

To receive initial equations we use the links between true σ_s and conditional σ_{so} stresses in form (1.42) which result from constant density condition. By applying these relations to initial rheological laws (2.39), (2.43) we obtain the expressions describing a development of strains in time as follows

$$\varepsilon_x S_y = \varepsilon_y S_x, \exp(-\alpha\varepsilon_{eq})\varepsilon_x = \Omega_1(t)(\sigma_{xo})^m D^{m-1} S_x \tag{6.30}$$

where $D = \sigma_{eq}/\sigma_{xo}$ and the values of S_x, S_y depend on the position of the symmetry axis relatively to the loading plane. For the cases when z, x, y are isotropy axes we have

$$S_x = \exp\varepsilon_x - kn_o \exp\varepsilon_y, \quad S_y = n_o \exp\varepsilon_y - k\exp\varepsilon_x,$$
$$S_x = (1-k)(2\exp\varepsilon_x - n_o\exp\varepsilon_y), \quad S_y = n_o\exp\varepsilon_y + (k-1)\exp\varepsilon_x,$$
$$S_x = \exp\varepsilon_x + (k-1)n_o\exp\varepsilon_y, S_y = (1-k)(2n_o\exp\varepsilon_y - \exp\varepsilon_x).$$

Here $n_o = \sigma_{yo}/\sigma_{xo}$.

Ultimate State

Any stage of the deformation can be taken as a limiting one but the most convenient is the use of criterion $d\varepsilon/dt \to \infty$ when at n_o = constant we find from (6.30) an equation for critical strains as

$$(\varepsilon_*)^2(1+R_*)n_o(1-k^2)\exp(\varepsilon_* + \varepsilon_{y*}) - C_* S_{x*}\varepsilon_* + (S_{x*})^2 = 0. \tag{6.31}$$

Here $\varepsilon_* = \varepsilon_{x*}$, $R = (m-1)\partial D/\partial\varepsilon_x + \alpha$ and functions $C(n_o, \varepsilon_x, \varepsilon_y)$ for the options above are respectively

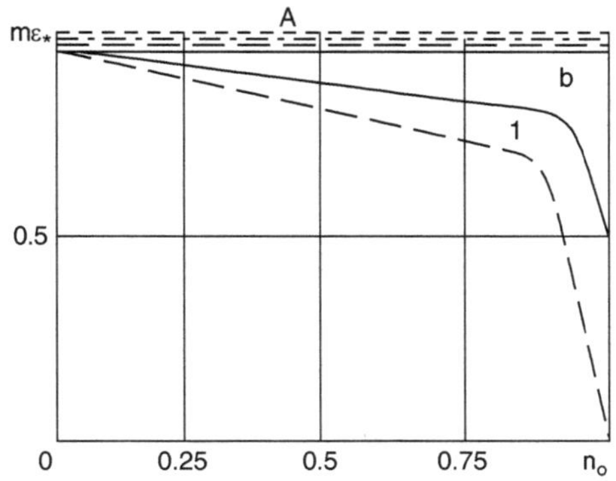

Fig. 6.13. Dependence of $m\varepsilon_*$ on m, k and n_o when z is axis of symmetry

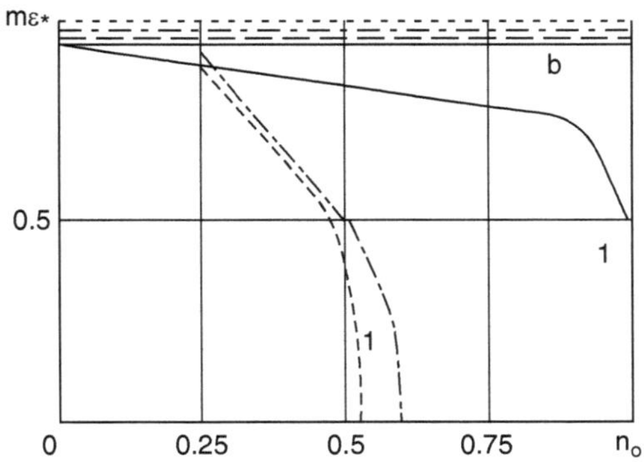

Fig. 6.14. Dependence of $m\varepsilon_*$ on m, k and n_o when y is axis of symmetry

$$
\begin{aligned}
C &= RS_x + \exp \varepsilon_x + n_o \exp \varepsilon_y, \\
C &= RS_x + 2(1-k) \exp \varepsilon_x + n_o \exp \varepsilon_y, \\
C &= RS_x + \exp \varepsilon_x + 2(1-k) n_o \exp \varepsilon_y. \qquad (6.32)
\end{aligned}
$$

From Figs. 6.13, 6.14 where for the cases when z and y are axes of symmetry (the case if x is axis of isotropy is similar of the first of them – this similarity was explained in Sect. 2.3.3) diagrams $m\varepsilon_*(m, k, n_o)$ for $\alpha = 0$, $m = 1$ and $m \to \infty$ (signs 1 and b) are represented we can see the great influence on critical state of the structure of m and k (solid, broken, interrupted by points and dotted lines for k-values equal to 0.5 (isotropic material – see Sect. 2.3.3), 1, 0, −1 respectively) and sign A means a validity for both m above.

Ultimate State of Plastic Materials

Expression (6.31) does not contain time and hence it can be used for plastic materials the limit state of which is usually evaluated by the dependence of critical sub-tangent Z_* /39/ (see Fig. 2.5)

$$Z_* = \sigma_{e*}(d\varepsilon_e/d\sigma_e)_*$$

to curve $\sigma_e(\varepsilon_e)$ on n. The transition to these variables can be made as

$$Z = m\varepsilon_e, n_o = n\exp(\varepsilon_x - \varepsilon_y) \tag{6.33}$$

and after transformations we obtain

$$(Z_*\sqrt{3}/2\sqrt{\Lambda_*})^2 n_*(1-k^2)(1+\alpha/m) - (Z_*\sqrt{3}/2\sqrt{\Lambda_*})C_{1*} + m = 0$$

where Λ are given by (2.40), (2.44) and values of $C_1(n)$ can be derived from (6.32) as follows

$$\begin{aligned} C_1 &= (m-1+\alpha)(1-kn)+1+n, \\ C_1 &= (m-1+\alpha)(1-k)(2-n)+2(1-k)n, \\ C_1 &= (m-1+\alpha)(1-(1-k)n)+1+2(1-k)n. \end{aligned} \tag{6.34}$$

The analysis shows (see Figs. 6.15, 6.16 in which diagrams $Z_*(n)$ are constructed for the same options and in the same manner as in Figs. 6.13 and 6.14) that parameters k, n, m have high influence on the ultimate plasticity of the body. Sometimes the same material can reveal absolute rigidity as well as infinite strains. We must notice that at $m \to \infty$, $\alpha = 0$ we have relation for Z_* in "classical" /39/ sense as

$$Z_* = 2\sqrt{\Lambda}/\sqrt{3}c_*$$

where for the options above

$$c = 1 - kn, \quad c = (1-k)(2-n), \quad c = 1 + (k-1)n.$$

For isotropic material at $k = 0.5$ and $\sigma_{eq} = 2\tau_e$ we receive

$$Z_* = 2\sqrt{1 - n_* + (n_*)^2}/(2 - n_*)\,.$$

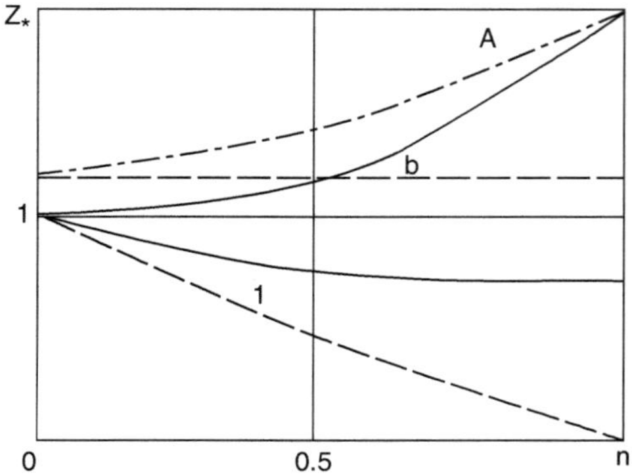

Fig. 6.15. Dependence $Z_*(n)$ when z is axis of symmetry

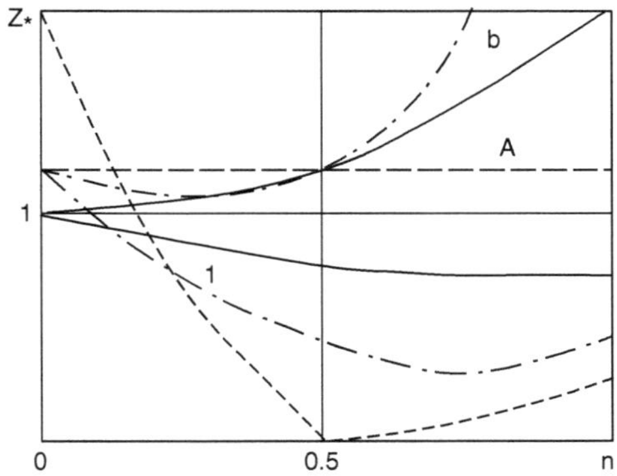

Fig. 6.16. Dependence $Z_*(n)$ when y is axis of symmetry

Ultimate State of Brittle Materials

When α is big we can neglect $m - 1$ and m in expressions (6.34) and for R_*. So, putting the second relation (6.33) into (6.31) we derive

$$\varepsilon_*^2 n(1 - k^2) - \varepsilon_* c^2 + c^2/\alpha = 0. \tag{6.35}$$

Solving this equation and decomposing the function under square root in series we have for two members of it a very simple and important result

$$\varepsilon_* = 1/\alpha \tag{6.36}$$

which coincides with the first relations (1.47), (5.129), (5.134) and others of that kind for all the structures in any stress state and can be considered as a "law of nature". Putting (6.36) into the second expression (6.30) we find the critical time

$$\Omega_1(t_*) = 1/\alpha e \sigma_{xo}^m D_*^{m-1} S_{x*}.$$

When the influence of time is negligible we deduce for $\Omega = $ constant

$$\sigma_{xo_*} = (\Omega_1 \alpha e D_*^{m-1} S_{x*})^\mu.$$

In the same manner the critical state of anisotropic thin-walled tube under axial load and internal pressure in three options of symmetry axis and loading plane as well as with addition of torsion (for case $k = 0{,}5$) can be studied /40/.(see also Appendix P).

Conclusion

In the book above we tried to put the bases for computations in the Geomechanics as a whole and the Soil Mechanics in particular for three types of body: elastic, plastic and hardening at creep with a damage. We demonstrated the methods of consequent sciences on the most important for practice examples (a slope under vertical loads, a compressed wedge, an action of a soil on a retaining wall, a flow of a material between two foundations, inclined plates and in a cone, a wedge and a cone penetration as well as the load-bearing capacity of piles and sheets of them, a tension and transversal shear of a body with a crack and a pressure on it of a punch, ultimate state of sphere, cylinder and cone under internal and external pressures, and plates at bending etc). Some of the problems were not studied due to the mathematical difficulties (e.g. a slope under combined loads for a non-linear material) or thanks to their unclear mechanical formulation.

We suppose as the most important Chaps. 5 and 6 that describe processes in structures between an initial elastic deformation and the final ultimate state, representing them as particular cases. To the regret Chap. 5 is the most difficult from mathematical point of view. However proposed here method of reducing the third order differential equations to the systems of the first order ones that can be solved one after another simplifies the procedure. We hope that this approach can be useful for some other tasks. We give also whenever is possible simple engineering relations.

The solutions in Chap. 6 take into account large displacements and strains that are often met in the geo-mechanical and other processes. Here the application of the criteria of infinite elongations and their rates is very convenient. The phenomenon of unstable change of strains due to a damage is also often met. The experiments show that the prediction of the method is nearer to a reality for more unsteady creep and non-linear link between stresses and strains. If we suppose that in the common sense the destruction is obliged to main elongations in dangerous body parts the method is widely used. Thanks to consideration of a damage (it is well-known that the theoretical strength of a material is much higher than a real one) this approach can be applied to structures that are destroyed at small change of dimensions and it was demonstrated in Chaps. 5 and 6. Here we underline as the most important result the independence of the maximum critical strain in brittle materials on the types of structure and stress state. In appendixes some auxiliary data are given. Among them the tables for simple computations of necessary values, details on calculations of some complex relations, a study of the fracture of brittle materials at eccentric compression, bases of applied creep theory similar to that in plasticity, the use of creep hypotheses for determination of the fracture time and others. Most results of the book have been received by the author in recent years and they are published in different editions throughout the world. (additionally to the author references above see also /41-47/) We hope that their compact presentation together with some well-known achievements in the field is necessary.

Herewith I use the occasion to remember the professor of Leningrad State University L.M. Kachanov who opened me the ideas of new approaches to Fracture Mechanics, and the professor of Odessa State University I.P. Zelinsky who put me for a solution some geo-technical problems. I also want to thank the workers of CIP Insel of Electro-technic faculty of Bochum University who helped me to create electronic version of this book.

References

1. Terzaghi K. Theoretical soil mechanics. N.Y. 1942, p. 510.
2. Frölich O.K. Bruchverteidigung in Baugrunde. Berlin. J. Springer. 1934.
3. Cytovich N.A. Soil Mechanics. 3rd edition. M. 1979, p. 272 (in Russian).
4. Sokolovski V.V. Static of soil media. London, Butterworth. 1960, p. 237.
5. Timoshenko S.P. Theory of elasticity. L-n – N.Y. 1934, p. 450.
6. Muschelisvili N.I. Some basic problems of the mathematical theory of elasticity. Noordhoff., 1975, p. 637.
7. Elsoufiev S.A. Modern methods of strength computations. Odessa, 1990, p. 60 (In Russian).
8. Griffith A. The phenomenon of rupture and flow of solids. Philosophical. Transactions of Royal Society, 1920 A-221, 163–198.
9. Gvozdev A.A. Development of the theory of reinforced concrete in the U.S.S.R., London, 1964.
10. Zelinski I.P., Elsoufiev S.A., Shkola A.V. Geomechanics. Odessa State University., 1998, p. 256 (in Russian).
11. Coulomb Ch. Essai sur une application des regle maximis et minimis a quelque problemes de static relatifs a l'architecture. Mem. Acad. Royal Pres. Divers Savants. Paris, 1776, v. 7.
12. Timoshenko S.P. History of strength of materials. N.Y., MxGraw-Hill Book Co., 1953, p. 225.
13. Hoff N. The necking and the rupture of rods subjected to constant tensile loads. J. Appl. Mech., 1953, v. 20, 105–108.
14. Elsoufiev S.A. The simplified rheological laws for the description of deformation and fracture of materials. Soviet Materials Science (USA). v. 19, 1. Translation from Physical-Chemical Mechanics of Materials 1982, 1, 72–76.
15. Carlsson R. Creep induced tensile instability. IJ of Mech. Sci. 1965, 7, 2, 218–222.
16. Elsoufiev S.A. To the solution of some non-linear problems in Geomechanics. Proceedings of 10th Danube-European Conference on SM&EF. 1995, 553–558.
17. Kachanov L.M. Introduction to continuum damage mechanics. Dordrecht, 1986, p. 280.
18. Sokolovski V. Theory of Plasticity. Moscow, 1969, p. 608 (In Russian).
19. Panasjuk V. Fracture mechanics. Successes and problems. Collection of abstracts of 8th IC on Fracture. Lviv (Ukraine), 1993, p. 3.

20. Sadowski N. Zeitschrift für Angewandte Mathematik und Mechanik. Bd.8, 1928, S. 107.
21. Rice J. Mathematical analysis in the mechanics of fracture. "Fracture". v. 2, Academic Press, N.Y., L-n, 1968.
22. Nadai R. Plasticity, 1931, p. 280.
23. Elsoufiev S.A. On bearing capacity of densified and reinforced grounds. Proceedings of the Third IC on Ground Improvement Geo-systems. L-n, 3–5 June 1997, 329–334.
24. Kachanov L.M. Foundations of the theory of plasticity. Amsterdam. 1971, p. 482.
25. Elsoufiev S.A. Ultimate state of a slope at non-linear unsteady creep and damage. In Proc. of IS on Slope Stability Engineering. A.A. Balkema, Rotterdam Brookfield, 1999, 213–217.
26. Elsoufiev S.A. Appreciation of load-bearing capacity of sluices' sliding supports, Friction and Wear (USA), 1985, 1, 171–174.
27. Elsoufiev S.A., Korneev A.B. Investigation of deformation characteristics of rings pile at cycling loading. Friction and Wear (USA), 1981, 161–164.
28. Hill R. Mathematical theory of plasticity, 1950, p. 407.
29. Elsoufiev S.A. An estimation of load-bearing capacity of some elements at their falling rigidity (USA). Translation from Machinovedenie, 1985, 6, 96–98.
30. Elsoufiev S.A. A scheme of determination of ultimate state of deformable bodies elements. Machinovedenie, 1981, 1, 65–69. (in Russian).
31. Elsoufiev S.A. Stability of slopes and cavities. In Proc. of GREEN 3 – The exploitation of material resources. Th. Tedford. L-n, 2001, 402–407.
32. Elsoufiev S.A. Computational method of ultimate state of structures' finding according to the method of transition to unstable condition. Machinovedenie, 1982, 5, 94–96 (in Russian).
33. Weil N. Tensile instability in thin-walled cylinders of finite length. J. of Mech. Sci. 1963, v. 5, 487–490.
34. Study of waterproofing revetments for the upstream face of concrete dams. Final report. GIGB, LCOLD, 1998, p. 164.
35. Weil N. Rupture characteristics of safety diaphragms. Transactions. of Am. Society. For Mech. Eng. (JAM), 1959, v. 4, 621–627.
36. Glebov V., Belyshev A. Experimental foundation of application of film membranes to computation. Izvestija USIG, 1976, v. 113, 65–69. (In Russian).
37. Natov N., Koleva V. Investigation of polymer plates at biaxial stress state. Mech. of composite materials., 1982, 3, 489–492. (In Russian).
38. Elsoufiev S.A. On one scheme of limit strains search at creep. Strength of Materials (USA). – Translation. from Problemy Prochnosti, 1978, 4, 86–89.
39. Lankford W., Saibel E. Some problems of unstable plastic flow under biaxial tension. Trans. of Amer. Institute of Mech. Engineers. Inst of Metal Division., 1947, 171, 562–571.
40. Elsoufiev S.A. Strength of anisotropic tubes at different loading. In Proc. of the Sixth European Mechanics of Materials Conference. Non-linear Mechanics of Anisotropic Mat., University of Liege/Belgium, Sept. 9–12, 2002, 227–232.
41. Elsoufiev S.A. On bearing capacity of piles and sheets of piles. In "Grouting Soil Improvement Geosystems Including Reinforcement." Helsinki, 2000, 457–465.
42. Elsoufiev S.A. Landslides due to propagation of cracks and plastic zones. In "Landslides in Research, Theory and Practice." Proc. of the 8th IS on Landslides, L-n, 2000, 507–512.

43. Elsoufiev S.A. Propagation of cracks and plastic zones at cycling loading. Extended Abstracts to Symposium on fatigue testing and analysis under variable amplitude loadings condition. Tours (France), 29–31 May 2002, 1–3 (No 68).
44. Elsoufiev S.A. Fracture of shells and membranes at non-linear unsteady creep and damage. In Abstracts of 76th Annual Scientific Conference of Gesellschaft für Angewandte Mathematik und Mechanik e.V. Part 2, 28 March-1 April 2005, 146–147.
45. Elsoufiev S.A. Rheology and fracture of composite materials. In book of Abstracts of IUTAM Symposium "Multiscale Modelling of Damage and Fracture Processes in Composite Materials." Kazimerz Dolny, Poland, 23–27 May 2005, p. 9.
46. Elsoufiev S. Fracture due to cracks and plastic zones near stamp edges propagation. In Abstracts of EC 466 "Computational and Experimental Mechanics of Advanced Materials." 20–22 July 2005. Loughborough, UK, p. 79.
47. Elsoufiev S.A. Tarasova B. English in texts on the Strength of Materials. Odessa, 1994, p. 92.
48. Il'jusin A.A. Plasticite. (Deformations elastico-plastique). Paris, Eyrolles, 1956.
49. Butin V., Elsoufiev S.A. Estimation of resistance to deformation of locks sliding supports. Trudy LIWT (Leningrad Institute of Water Transport) No 158. Waterways and hydro-technical structures. Leningrad, 1977, 156–163 (In Russian).
50. Nikiforov N.N.,Chebanov V.M., Elsoufiev S.A. Method of polymer materials tests to gas penetration at high pressure. Work on call.1972, No 125, 38–47, CNIITC (in Russian).
51. Elsoufiev S.A., Shatalin K.I. Theory of Elasticity and Plasticity, Leningrad, 1982, p. 79. (In Russian).
52. Barrer R. Diffusion in and through solids. 2 impr. Cambridge Univers. Pr. 1951.
53. Nowoshilov V. On the link of stresses and strains in non-linear elastic body. Appl. Math. and Mech., 1951, v. 15, 2 (in Russian).

Appendices

A

Computation of p* for Brittle Materials Which do not Resist Tension

In Germany the computation of ultimate load for brittle material is usually made according to the schemes that are represented in Fig. A.1.

If $e = h/6$ force F is in the edge of the core /47/ (Fig. A.1,a) and the maximum stress is

$$p^* = 2F/h$$

and F is the first ultimate load

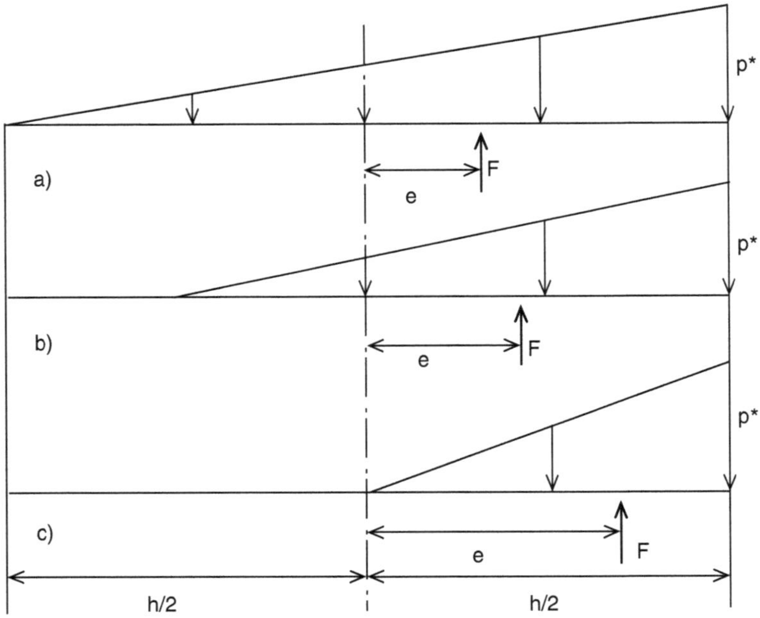

Fig. A.1. Distribution of stresses at eccentric compression in brittle materials like soils

If $h/6 < e < h/3$ (the force is out of the core – Fig. A.1,b) we receive from static equation

$$p^* = 4F/3(h - 2e).$$

If $e = h/3$ (damage reaches the axis of symmetry – Fig. A.1,b) we have

$$p^* = 4F/h$$

and F is the second ultimate load which in this case is less than the first one.

B

Values of K_σ in (3.127)

Values of coefficients for computation of compressive stresses due to concentrated force

r/z	K	r/z	K	r/z	K	r/z	K	r/z	K	r/z	K
0.00	0.4775	0.32	0.3742	0.64	0.2024	0.96	0.0933	1.28	0.0422	1.60	0.0200
0.02	0.4770	0.34	0.2632	0.66	0.1934	0.98	0.0887	1.30	0.0402	1.62	0.0191
0.04	0.4756	0.36	0.3521	0.68	0.1846	1.00	0.0844	1.32	0.0383	1.64	0.0183
0.06	0.4732	0.38	0.3408	0.70	0.1762	1.02	0.0803	1.34	0.0365	1.66	0.0175
0.08	0.4699	0.40	0.3294	0.72	0.1684	1.04	0.0764	1.36	0.0348	1.68	0.0167
0.10	0.4657	0.42	0.3181	0.74	0.1603	1.06	0.0727	1.38	0.0332	1.70	0.0160
0.12	0.4607	0.44	0.3068	0.76	0.1527	1.08	0.0694	1.40	0.0317	1.72	0.0153
0.14	0.4548	0.46	0.2955	0.78	0.1455	1.10	0.0658	1.42	0.0302	1.74	0.0147
0.16	0.4482	0.48	0.2843	0.80	0.1398	1.12	0.0626	1.44	0.0288	1.76	0.0141
0.18	0.4409	0.50	0.2733	0.82	0.1320	1.14	0.0595	1.46	0.0275	1.78	0.0135
0.20	0.4329	0.52	0.2625	0.84	0.1267	1.16	0.0567	1.48	0.0263	1.80	0.0129
0.22	0.4242	0.54	0.2518	0.86	0.1196	1.18	0.0539	1.50	0.0251	1.82	0.0124
0.24	0.4151	0.56	0.2414	0.88	0.1138	1.20	0.0513	1.52	0.0340	1.84	0.0119
0.26	0.4054	0.58	0.2313	0.90	0.1083	1.22	0.0489	1.54	0.0229	1.86	0.0114
0.28	0.3951	0.60	0.2214	0.92	0.1031	1.24	0.0466	1.56	0.0219	1.88	0.0109
0.30	0.3849	0.62	0.2117	0.94	0.0981	1.26	0.0443	1.58	0.0209	1.90	0.0105

C

Values of K′ in (3.128)

Values of coefficients K′ for computation of stresses by the method of corner points

n =	m = a/b											
z/b	1.0	1.2	1.4	1.6	1.8	2.0	2.2	2.4	2.6	3.0	4.0	8.0
0.0	0.250	0.250	0.250	0.250	0.250	0.250	0.250	0.250	0.250	0.250	0.250	0.250
0.2	0.249	0.249	0.249	0.249	0.249	0.249	0.249	0.249	0.249	0.249	0.249	0.249
0.4	0.240	0.242	0.243	0.243	0.244	0.244	0.244	0.244	0.244	0.244	0.244	0.244
0.6	0.223	0.228	0.230	0.232	0.232	0.232	0.233	0.233	0.234	0.234	0.234	0.234
0.8	0.200	0.202	0.212	0.215	0.215	0.217	0.218	0.219	0.219	0.220	0.220	0.220
1.0	0.175	0.185	0.191	0.196	0.198	0.200	0.201	0.202	0.203	0.203	0.204	0.205
1.2	0.152	0.163	0.171	0.176	0.179	0.182	0.184	0.185	0.186	0.187	0.188	0.189
1.4	0.131	0.142	0.153	0.157	0.161	0.164	0.167	0.169	0.170	0.171	0.173	0.174
1.6	0.112	0.124	0.133	0.140	0.145	0.148	0.151	0.153	0.155	0.157	0.159	0.160
1.8	0.100	0.108	0.117	0.124	0.129	0.133	0.137	0.139	0.141	0.143	0.146	0.148
2.0	0.084	0.095	0.103	0.110	0.116	0.120	0.124	0.126	0.128	0.131	0.135	0.137
3.0	0.045	0.052	0.058	0.064	0.069	0.073	0.078	0.080	0.083	0.087	0.093	0.098
4.0	0.027	0.032	0.036	0.040	0.044	0.047	0.051	0.054	0.056	0.060	0.067	0.075
5.0	0.018	0.021	0.024	0.027	0.030	0.033	0.036	0.038	0.040	0.044	0.050	0.060
7.0	0.009	0.011	0.013	0.015	0.016	0.018	0.020	0.021	0.022	0.025	0.031	0.041
10.0	0.005	0.006	0.007	0.007	0.008	0.009	0.010	0.011	0.012	0.013	0.017	0.026

D

Values of β_1, β_2 in Fig 4.33 and factors A, B in (4.88)

Table D.1. Data for construction of centre of the most dangerous slip arc when soil has only coherence

Gradient of slope	β_1	β_2
	In degrees	
1.73:1	29	40
1:1	28	37
1:1.5	26	35
1:2	25	35
1:3	25	35
1:5	25	37

Table D.2. Values of coefficients A, B for approximate computation of slope stability

Gradient of slope	Slip surface goes through lower edge of slope		Slip surface goes through base of slope and has horizontal tangent in depth							
1:m			l = h/4		l = h/2		l = h		l = 1.5h	
	A	B	A	B	A	B	A	B	A	B
1:1.00	2.34	5.79	2.56	6.10	3.17	5.92	4.32	5.80	5.78	5.75
1:1.25	2.64	6.05	2.66	6.32	3.24	6.62	4.43	5.86	5.86	5.80
1:1.50	2.64	6.50	2.80	6.53	3.32	6.13	5.54	5.93	5.94	5.85
1:1.75	2.87	6.58	2.93	6.72	3.41	6.26	4.66	6.00	6.02	5.90
1:2.00	3.23	6.70	3.10	6.87	3.53	6.40	4.78	6.08	6.10	5.95
1:2.25	3.18	7.27	3.26	7.23	3.63	6.56	4.90	6.16	6.18	5.98
1:2.50	3.53	7.30	3.46	7.62	3.82	6.74	5.08	6.26	6.26	6.02
1:2.75	3.59	8.02	3.68	8.00	4.02	6.95	5.17	6.36	6.34	6.05
1:3.00	3.59	8.91	3.93	8.40	4.24	7.20	5.31	6.47	6.44	6.09

E

Flow of Ideal Plastic Material in Cone

The vicinity of precise and approximate dependences (Fig. 4.41) allows to use for practical purposes rigorous solution (4.104). Putting it into (4.105) we find

$$\sigma_\chi = \tau_{yi}(2n\ln(a/\rho) - J/\sin^2\lambda - 3J_o(\chi))$$

where

$$J_o(\chi) = \int_0^\chi \sin 2\psi d\chi$$

Diagram $J_o(\lambda)$ is given in Fig. E.1. Jntegral J has the same meaning as in relation (4.106) but it must be computed so:

$$J = \int_0^\lambda (2\cos 2\psi - 3J_o(\chi))\sin 2\chi d\chi. \tag{E.1}$$

Curve $J(\lambda)$ according to (E.1) is represented by broken line in Fig. 4.42 and we can see that it is not far away from solid one. Consequently dependence $\max \tau_e/q$ in Fig. 5.22 is near the dotted line according to relation (4.106).

In the similar way we compute instead of (4.108)

$$P_*/\pi = (a+1)^2(p* - 2n\tau_{yi}\ln(1+1/a)\sin^2\lambda) + \tau_{yi}(1(1+2a)(\cos\lambda\sin\lambda + J_2/4 + ((n - 3J_o(\lambda))\sin^2\lambda)$$

where (at n = 2.045 e.g. $J_2 = 6$)

$$J_2 = \int_0^\lambda \Big((2\cos 2\psi - 3J_o(\chi))\cos\chi + \sin 2\psi \sin\chi\Big)\sin\chi d\chi$$

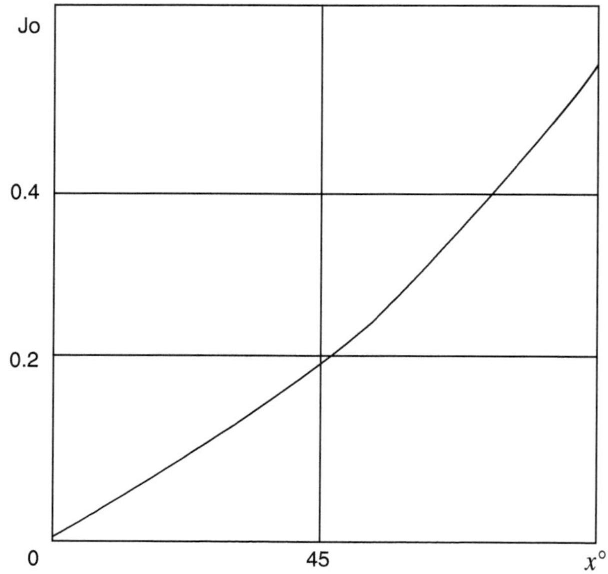

Fig. E.1. Dependence $J_o(\chi)$

F

Computation of Stresses at Anti-plane Deformation of Massif with Crack and Moving Punch

We put in (5.6) $C - \theta = x$ and rewrite solution of (5.4) as

$$4\ln(f/D) = \pm((m-1)/(m+1))\ln((\surd - (m-1)\cos x)/(\surd + (m-1)\cos x))$$
$$+((\surd + (m+1)\cos x)/(\surd - (m+1)\cos x)) + \ln\sin^2 x \qquad (F.1)$$

where

$$\surd = \sqrt{(m+1)^2 - (m-1)^2\sin^2 x}.$$

At x near zero we have from (F.1) at lower and (5.4) at upper signs

$$f = D_1(1/m)^{(m-1)/4(m+1)}\sqrt{2}\cos((C-\theta)/2),\ f'$$
$$= \left(D_1/\sqrt{2}\right)(1/m)^{(m-1)/4(m+1)}\sin((C-\theta)/2) \qquad (F.2)$$

Border demands for crack $f'(0) = 0$, $f(0) = 1$ give $C = 0$ and value D_1(5.7). Condition $f(\pi) = 0$ is fulfilled and from the second (F.2) we have

$$/f'(\pi)/ = D_1^2.$$

At x near zero we have from (F.1), (5.4) at upper and lower signs respectively

$$f = D_2 m^{(m-1)/4(m+1)}\sqrt{2}\sin(C-\theta)/2),\ f'(0)$$
$$= (D_2/\sqrt{2}m^{(m-1)/4(m+1)}\cos((C-\theta)/2). \qquad (F.3)$$

The conditions $f'(\pi) = 0$, $f(\pi) = 1$ for the punch give $C = 0$,

$$D_2 = (1/m)^{(m-1)/4/m+1)}/\sqrt{2} \qquad (F.4)$$

and $f(0) = 0$, $f'(0) = D_2^2$.

Now we rewrite (5.8) for the crack as following

$$R_1(\theta) = \left(\sqrt{(mf/(m+1))^2 + f'^2}\right)^{m+1} /I(m)/f'(\pi)/^m \qquad (F.5)$$

Table F.1. R_1, R_2 values

m	7	3	1	3	7
θ	crack			punch	
0	0.04	0–34	0.32	0.12	0.03
π/4	0.13	0.38	0.32	0.60	0.12
π/2	0.08	0.26	0.32	0.45	5.86
3π/4	0.11	0.17	0.32	0.76	1.36
π	0.15	0.20	0.32	0.25	0.20

where

$$R_1 = 2G\Omega(t)(\tau_e)^{m+1}r/\pi\tau_o^2 l. \tag{F.6}$$

Relation (5.5) may be applied to the punch with replacement of $f'(\pi)$ by $f'(0)$ with

$$R_2 = 2G\Omega r\tau_e^{m+1}/\pi l Q^{2.}$$

The values of R_1, R_2 are given in the Table F.1 above and as we can see from it the plastic zones may be big. Because of that we study them for one-sided crack Fig. 5.2, b) in elastic-plastic material with hardening (see diagram $\tau = \tau(\gamma)$ in Fig. 5.2.a) where $\tau \equiv \tau_e$, in (2.52) and $\gamma \equiv \gamma_m$ in (2.54) as well as

$$\tau_x = \tau(\gamma)\gamma_x/\gamma, \tau_y = \tau(\gamma)\gamma_y/\gamma. \tag{F.7}$$

Instead of static (2.48) and compatibility (2.50) equations Rice /21/ used reversed laws

$$\partial x/\partial\tau_x + \partial y/\partial\tau_y = 0, \partial x/\partial\gamma_y - \partial y/\partial\gamma_x = 0. \tag{F.8}$$

The second of these relations may be satisfied by the introduction of potential function ψ as

$$x = \partial\psi/\partial\gamma_x, y = \partial\psi/\partial\gamma_y. \tag{F.9}$$

Now we consider in plane γ_x,γ_y polar coordinates γ, φ where φ is calculated from vertical axis γ_y. Then $\psi = \psi(\gamma,\varphi)$ and using (F.9) as well as expressions of transformation from γ_x, γ_y to γ, φ (in /51/ the simplified method of it is represented) we receive

$$x = -\sin\varphi\partial\psi/\partial\gamma - \cos\varphi\partial\psi/\gamma\partial\varphi, y = \cos\varphi\partial\psi/\partial\gamma - \sin\varphi\partial\psi/\gamma\partial\varphi. \tag{F.10}$$

Putting these relations in the first (F.8) we obtain differential equation

$$(\tau(\gamma)/\tau'(\gamma))\partial^2\psi/\partial\gamma^2 + \partial\psi/\partial\gamma + \partial^2\psi/\gamma\partial\varphi^2 = 0. \tag{F.11}$$

We shall seek solution of (F.11) in form

$$\psi = f(\gamma)\sin\varphi \tag{F.12}$$

which gives linear differential equation of the second order for $f(\gamma)$. At boundary conditions $f'(\infty) = 0$, $f(\gamma_{yi}) = 1$ and the rheological equation in Fig. 5.2a, it gives the solution

$$\psi = (\gamma_{yi}/\gamma)^{\mu} \sin\varphi \tag{F.13}$$

Putting (F.13) into (F.10) with border demand $x = -l$ at $\varphi = -\pi/2$ we find the boundary of plastic zone ($\gamma = \gamma_{yi}$)

$$x/l = m\cos^2\varphi - \sin^2\varphi, y/l = -0.5(\mu+1)m\sin 2\varphi$$

which for $m = 1, 3, 7$ is given by solid, broken and interrupted by points curves in Fig. 5.2b. (upper part must be constructed according to symmetry). We can see that with a growth of m the plastic zone increases and moves out of the crack.

G

Some Computations on Bending of Wedge

Some materials (including soils) have stress-strain diagram with $\mu > 1$. We consider the case $\mu = 2$ when from (5.71) and boundary conditions we have

$$g \cos 2\psi = H \sin \theta.$$

Then from (5.70) we derive differential equation

$$d\psi/d\theta = \sin(\theta - 2\psi) \cos 2\psi/(1 + \sin^2 2\psi) \sin \theta. \quad \text{(G.1)}$$

We integrate (G.1) at different $\Theta(0) \equiv \Theta_o$ by the finite differences method. The diagrams $\psi(\theta)$ for $\Theta_o = 1, 2, 3, 4$ are given in Fig. G.1 by curves 1, 2, 3, 4 respectively. The consequent values of λ are 40°, 64°, 87° and 117°.

Then according to (5.69) we have

$$\tau = (\omega H^2/r^2))(\sin \theta/\cos 2\psi)^2 \sin 2\psi$$

and with consideration of (5.72) and (G.1) we compute

$$\max \tau_e = M\Theta_o^2/2Jr^2$$

where

$$J = \int_0^\lambda (\sin \theta/\cos 2\psi)^2 \sin 2\psi d\theta.$$

The diagram $\max \tau_e r^2/M$ as a function of λ is given in Fig. 5.13 by interrupted by points line and we can see that it is much higher than that one for $m = 1$.

As particular case of the general theory we take it for $m = 2$ when (5.78) becomes

$$d\Theta/d\psi = -2\Theta(1 - \Theta)(1 + 2(1 - \Theta)/\Psi) \quad \text{(G.2)}$$

where $\Psi = (1 + \cos^2 2\psi)/2$. We integrate this equation at Θ_o equal to 0.5 and 1.249, and draw the curves $\psi(\theta)$ in Fig. G.2 with numbers 0.5 and 1.25 (for the case $\Theta_o = \Theta = 1$ we have straight line 1 with relation $\psi = \theta - \pi/4$).

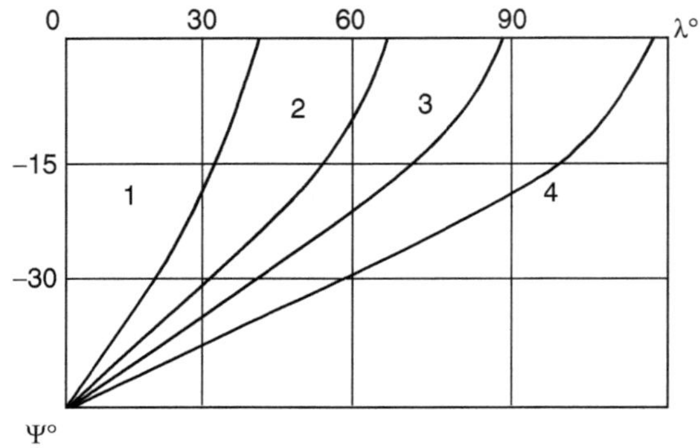

Fig. G.1. Diagram $\psi(\lambda)$ for $\mu = 2$

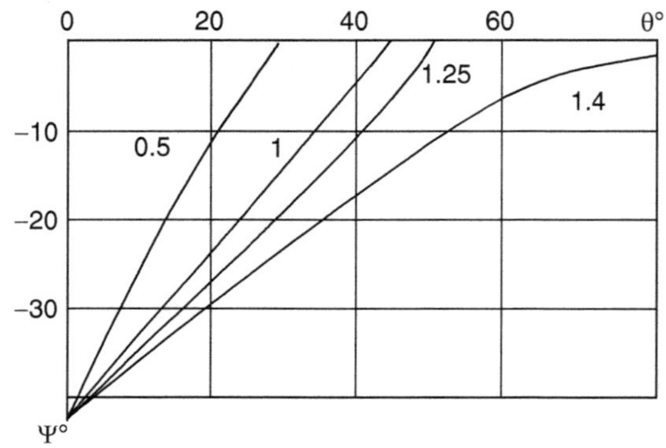

Fig. G.2. Curves $\psi(0)$ for case $\mu = 0.5$

For $\Theta_o > 1.25$ we can not find λ since at small ψ $\theta \to \infty$ (e.g. curve 1.4 for $\Theta_o = 1.4$ in the figure). Computing by (5.82) M-value we obtain $\max \tau_e$ according to (5.81) which is shown by pointed line in figure 5.13, and we can see that it is near the solid curve (for $m = 1$) and, so, for $\Theta_o > 1.25$ the latter must be used as the first approach (similar to the case in Sect. 5.2.2).

In conclusion we derive similar to Sect. 5.2.2 an engineering solution. We take with consideration of (5.66) rheological law like (5.30)

$$\varepsilon_\theta = -3\Omega(t) r^{-2m} F^{(m-1)/2} f'/4, \ \gamma = 3\Omega(t) r^{-2m} F^{(m-1)/2} f$$

where $F = f'^2 + 4f^2$ and put it into (2.71) which leads to non-linear differential equation

$$\begin{aligned}(m-1)\left((m-3)\left(f''+4f\right)^2 f'^2 + F\left(f'''+4f'\right)f'^2 + 3F\left(f''+4f\right)f'f'' - 4mF^2f'\right)\\ + F^2f''' + 4\left(2m-1\right)\left((m-1)F\left(f''+4f\right)ff' + F^2f'\right) = 0.\end{aligned}$$

At $m = 1$ it gives the solution of Sect. 5.2.5. If we suppose $f' = 0$ in the whole wedge we derive simple expression $f''' = 0$ which with consideration of boundary conditions (5.72) and static demand (5.72) provides solution

$$f = 3M(\theta^2 - \lambda^2)/4\lambda^3.$$

The value of $\max \tau_e$ takes place at $\theta = 0$ and is

$$\max \tau_e = 3M/4r^2\lambda. \tag{G.3}$$

The computations show that at $\pi/6 < \lambda < \pi/4$ $\max \tau_e$ is near the curves for $\mu = 0$, $\mu = 0.5$, $\mu = 1$ in Fig. 5.13 and at $\pi/4 < \lambda < \pi/2$ it is somewhat below of the solid line ($\mu = 1$). So, in these limits engineering relation (G.3) can be used as the first approach.

H

Bases of Applied Creep Theory

As was shown in Sect. 4.2.2 a tension or a compression of a thin layer induces at plane deformation almost equal triple-axial stress state with small shearing components. It opens the way to the construction of the applied plasticity theory /48/. In a similar manner the applied creep theory can be formulated.

We create it for the part I in Fig. 5.14. Since σ_x does not depend on y we write down the first equation (2.59) at X = 0 as

$$d\sigma_x/dx - 2\tau/h = 0$$

or after the use of the Coulomb's law (4.44)

$$d\sigma_x/dx = -2f\sigma_y/h. \tag{H.1}$$

As was shown in /49/ strain ε_x is a constant (its dependence on time is further hinted) and according to (5.84) σ_x, σ_y are functions of x only. Putting (5.84) into (H.1) and integrating we receive

$$\sigma_y = (\sigma_o - 4\omega(t)(\varepsilon_x)^\mu)e^{2f(a-x)/h} \tag{H.2}$$

Now we find the specific pressure in y direction as following

$$N/2a = (1/a)\int_0^a \sigma_y\,(x)\,dx.$$

Replacing here σ_y from (H.2) we have after integration relation

$$N/2a = (\sigma_o - 4\omega(t)(\varepsilon_x)^\mu)h(e^{2fa/h} - 1)/2fa \tag{H.3}$$

from which the first expression (5.85) is received with compressing force N taken according to its absolute value. To obtain the second law (5.85) (for part II in figure 5.14) we must replace in (H.3) N/2a, ε_x, σ_o h, 2a by σ_o, ε_y, 0, (b-a), h respectively.

In the same manner the case of material flow between slightly inclined plates can be considered (/10, 49/).

The solutions in Sects. 3.2.2 and 4.2.2 (see Fig. 3.10) can be generalized for pressed thin layer of hardening at creep material. From hypothesis of plane cross-sections we have $u_y = u_y(y)$. The incompressibility condition and (2.60) give with a consideration of symmetry demand

$$u_x = -u_y'(y)x, \ \gamma = -u_y''(y)x \tag{H.4}$$

From relations (5.34) and (5.17) with replacement in them r, θ by x, y we have

$$\sigma_x = \sigma_y, \ \tau = -\omega(t)(u_y'')^{\mu}x^{\mu} \tag{H.5}$$

and for thin layer

$$\tau = -A\omega(t)yx^{\mu}. \tag{H.6}$$

where A is constant. Now we receive from static relation (2.59) at X = 0

$$\sigma_x = C - A\omega(t)x^{\mu+1}/(\mu + 1) \tag{H.7}$$

where C may be found from integral equilibrium equation

$$\int_{-h}^{h} \sigma_x (l, y)\, dy = 0$$

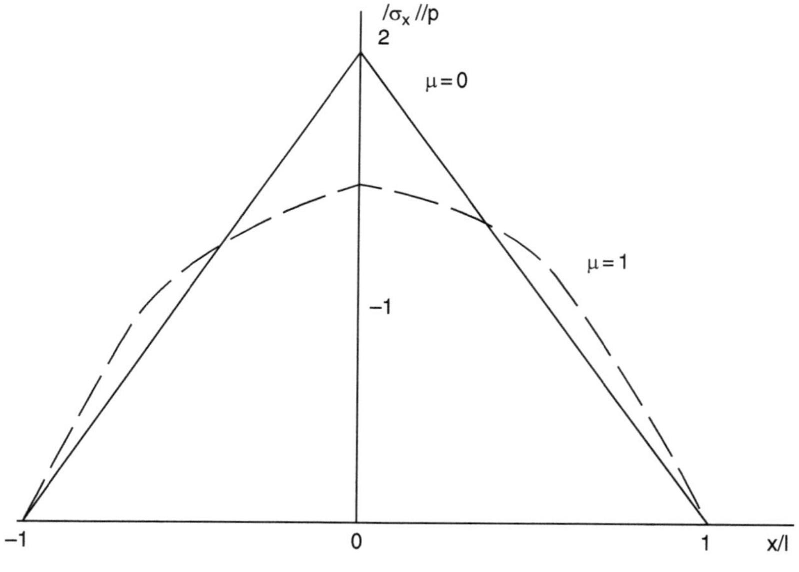

Fig. H.1. Distribution of normal stresses in thin layer

which gives with consideration of the first (H.5)

$$\sigma_x = \sigma_y = A\omega(t)(x^{\mu+1} - l^{\mu+1})/(\mu + 1). \tag{H.8}$$

Another similar demand

$$\int_0^l \sigma_y\,(x)\,dx = -pl$$

(p = P/2l) reduces to

$$A\omega(t)l^{\mu+1} = (\mu + 2)p$$

and hence instead of (H.5), (H.6) we have respectively

$$\tau_{xy} = -(\mu + 2)p/x/l/^{\mu}y/l,\ \sigma_x = (\mu + 2)p(/x/l/^{\mu+1} - 1)/(\mu + 1). \tag{H.9}$$

For $\mu = 1$, $\mu = 0$ we receive solutions in Sects. 3.2.2 and 4.2.2 at $y = h$ and $\sigma_x(0, h) = \sigma_y(0, h) = 0$.

Plane deformation ($\varepsilon_z = 0$) gives $\sigma_z = \sigma_x = \sigma_y$ and since for thin layer $h \ll l$ a material is in triaxial equal tension or compression under small shearing stresses and to it the conclusion of Sect. 4.2.2 may be applied. We can also see from Fig. H.1 that with growth of μ the distribution of normal stresses becomes more even.

I

Inelastic Zones Near Crack in Massif at Tension and Pressed Punch

We rewrite (5.99) as

$$R = F^{m+1}/I_1(m) \tag{I.1}$$

where for the crack at plane deformation (k = 1)

$$R = R_1 = r(2\tau_e)^{m+1}G\Omega_1(t)/\tau^2\pi l,$$

Using border demands in Sect. 5.2.7 we derive for m = 3, 15 respectively

$$f = 0.383\cos 0.4\theta + 0.572\sin 0.4\theta + 0.616\cos 0.915\theta - 0.25\sin 0.915\theta,$$
$$f = -0.4616\cos 0.232\theta + 1432\sin 0.232\theta + 5.616\cos 0.386\theta - 8.62\sin 0.386\theta.$$

Putting $f'''(\pi)$. $f''(\pi)$ into expression for $I_1(m)$ in Sect. 5.2.7 we find $I_1(3) = 0.14$, $I_1(15) = 0.033$. According to (I.1) and (5.94) we compute R_1 for some θ (left part of Table I.1 and Fig. I.1)

Table I.1. Values of R_1, R_2

m	15	3	1	3	7	15
θ		crack			punch	
0	0.0	0	0	0.200	0.022	0.002
$\pi/4$	0.058	0.840	0.08	0.857	0.471	0.097
$\pi/2$	3.197	2.755	0.16	1.843	2.481	2.365
$3\pi/4$	0.142	0.792	0.08	0.604	1.598	0.040
π	0.002	0.031	0	0	0	0

In similar way we use (I.1) for the punch when at

$$R = R_2 = r(2\tau_e)^{m+1}\Omega\pi Gl/P^2$$

at m = 3, m = 7, m = 15 we have

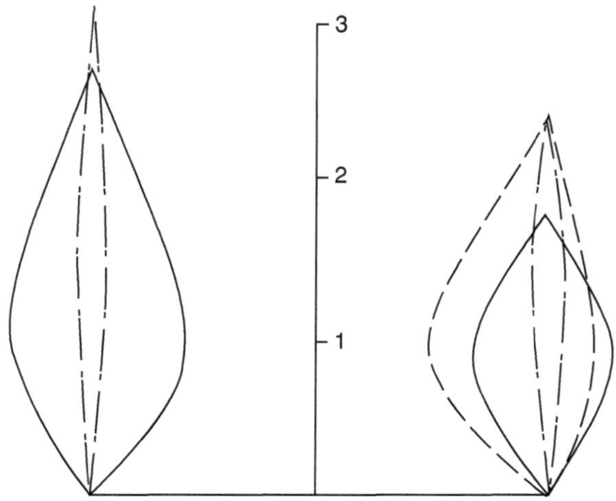

Fig. I.1. Inelastic Zones Near Edges of Crack and Punch

$$f = -0.665\cos 0.4\theta - 0.183\sin 0.4\theta + 0.665\cos 0.915\theta + 0.08\sin 0.915\theta.$$

$$f = -2.266\cos 0.31\theta + 2.75\sin 0.31\theta + 2.266\cos 0.584\theta - 1.46\sin 0.584\theta.$$

$$f = -4.616\cos 0.232\theta + 14.32\sin\theta + 5,616\cos 0.386\theta - 8.62\sin 0.386\theta$$

Putting $f'''(0)$, $f''(0)$ instead of $f'''(\pi)$, $f''(\pi)$ in expression for $I_1(m)$ at $I_1(3) = 0.206$, $I_1(7) = 3$. $I_1(15) = 0.06$ we find R_2 at the same θ as before are given in Table and Fig. I.1 with solid, broken and interrapted by points line). The values of R for $m = 1$ coincide and it is equal

$$R(1) = (\sin^2\theta)/2\pi.$$

From Table I.1 and Fig. I.1 we can see that with a growth of m the inelasic zone changes its form and dimensions.

J

Inelastic Zones Near Crack and Punch Ends at Transversal Shear

We rewrite (5.100) as

$$R(\theta) = F^{m+1}/I(m)/f'(\pi)/^{m} \tag{J.1}$$

where for the plane deformation

$$R = 4r(2\tau_e)^{m+1}(G\Omega(t)/\tau^2)\pi.$$

At $m = 1$ we have from (J.1) relation

$$R = (1 + 3\cos^2\theta)/2\pi \tag{J.2}$$

which coincides with equation for τ_e from (3.106).

For $m = 3$ we have $I(3) = 3.33$ and

$$f = 0.997\cos 0.342\theta + 0.555\sin 0.342\theta - 0.997\cos 1.873\theta + 0.461\sin 1.873\theta. \tag{J.3}$$

Then we find $f'(\theta), f''(\theta), /f''(\pi)/^3 = 29.28$ and R (left part of Table J.1).

For $m = 15$ according to boundary conditions we detail (5.96) as

$$f = -1.04\cos 0.162\theta - 0.515\sin 0.162\theta + 1.04\cos 2.146\theta - 0.505\sin 2.146\theta.$$

Then we compute $f'(\theta), f''(\theta)$ and $/f''(\pi)/^{15} = 7.14 \times 10^{10}$, $I(15) = 15$. The values of R are given in Table J.1 and we can see that all the curves have similar form. The computations show that the stresses are near to consequent curve 1 in Fig. 3.21.

Relation (J.1) may be used to finding inelastic zones near punch edges with $f''(0)$ instead of $f''(\pi)$ and

$$R = 2r(2\tau_e)^{m+1}\pi G\Omega_1 l/Q^2.$$

The computations show that the results are identical to the data in Table J.1 reversed about axis $\theta = \pi/2$.

Table J.1. Values of R(θ) at different m

m	1	3	15
θ		Crack	
0	0.64	1.24	0.20
π/4	0.40	0.52	0.10
π/2	0.32	0.38	0.17
3π/4	0.40	0.61	0.04
π	0.64	0.93	0.35

K

Flow of Material in Cone

In order to demonstrate the computations in general case we make them for the option $\mu = 1/3$ when equation (5.119) becomes

$$\begin{aligned} d\Theta/d\chi = {} & \Theta/\tan\chi - (4/3)\tan 2\psi + (4 + \Theta(1.5\Theta - 5))(\tan 2\psi)/3\Psi \\ & - \Theta(4 + 0.75\Theta(\tan 2\psi/\tan\chi + 1 + 1/\sin^2\chi))\sin 4\psi/3\Psi \end{aligned} \tag{K.1}$$

where according to (5.49)

$$3\Psi = 1 + 2\cos^2 2\psi.$$

We integrate (K.1) at different $\Theta(0) = \Theta_o$ and (5.118) at boundary condition $\psi(0) = 0$ by the finite differences method. The curves $\psi(\chi)$ at $\Theta_o = 1, 2, 4$ are marked in Fig. K.1 by numbers 1, 2, 3 respectively.

Then we integrate the second expression (5.113) as

$$g = \exp\left(2\int_0^\chi (1/\Theta - 1.5)\tan 2\psi d\chi\right)$$

neglecting an arbitrary multiplier because it is in numerator as well as in denominator of (5.120). Lastly we compute J_3 and use (5.120). The diagrams of $\max\tau_e(\lambda)$ is given in Fig. 5.22 by interrupted by points line and we can see that it is near to that one at $\mu = 2/3$.

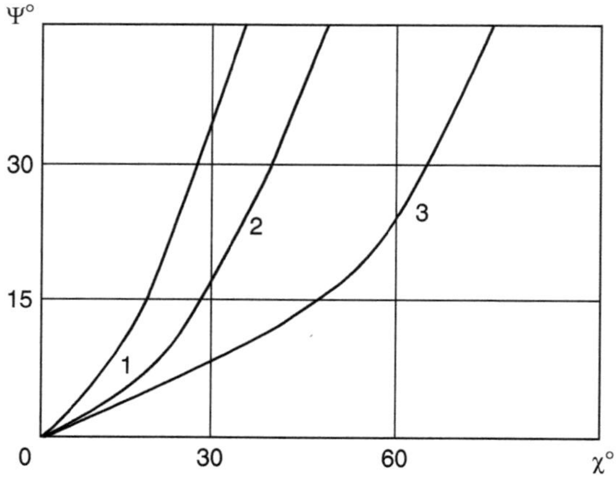

Fig. K.1. Diagrams $\psi(\chi)$

L

The Use of Hypotheses of Creep

We shall apply them to the case when (see sub-paragraph 1.5.5)

$$\varepsilon = B\sigma^m t^n \quad (0 \leq n \leq 1). \tag{L.1}$$

The validity of this expression at $\sigma \neq$ constant represents a time hardening theory which is mainly used in this book. Differentiation of (L.1) at $\sigma =$ constant gives the flow hypothesis

$$d\varepsilon/dt = Bn\sigma^m t^{n-1}. \tag{L.2}$$

Combination of (L.1) and (L.2) allows to formulate strain hardening theory in relation

$$\varepsilon^{1/n-1} d\varepsilon/dt = nB^{1/n}\sigma^{m/n}. \tag{L.3}$$

The well-known heredity integral must be written as following:

$$\varepsilon = Bn \int_0^1 \sigma^m(\xi)(t-\xi)^{n-1} d\xi \tag{L.4}$$

At $\sigma =$ constant equations (L.2)–(L.4) coincide with (L.1).

To appreciate the validity of these hypotheses we must use them for the regimes with $\sigma \neq$ constant e.g. $\sigma = ut$ where $u =$ constant. In this case we have a common expression

$$\varepsilon = KB\sigma^m t^n \tag{L.5}$$

where K is equal to 1, $n/(m+n)$, $(n/(m+n))^n$, $\Gamma(m+1)\Gamma(n+1)/\Gamma(m+n+1)$ for relations (L.1), (L.2), (L.3), (L.4) respectively. As for non-linear unsteady creep m is big and n – small the flow hypothesis predicts negligible K-values against other theories, e.g. for $m = 3$, $n = 1/8$ we receive $K = 0.04$, 0.67 and 0.8. The experiments /14/ give as a rule data between the predictions of strain hardening and heredity hypotheses and in this situation a special meaning receives the time hardening theory with its advantage of simplicity for practical applications.

To use the Hoff's method we put in (L.2), (L.3) expression (1.42), (1.43) and integrate in limits 0, t_∞ and 0, ∞ for t, ε that gives respectively

$$t_\infty = (mB(\sigma_o)^m)^{-1/n}, \; t_\infty = n^{1-1/n}\Gamma(1/n)(B\sigma_o^m m)^{-1/n}. \qquad (L.6)$$

At $n = 1$ (a steady creep) they coincide with (1.44). At $n < 1$ the first of them leads to bigger values of the fracture time than the second one.

Lastly, we compute the fracture and critical times according to strain hardening theory and (1.46) at $\alpha = 0$, $\Omega(t) = Bt^n$ as

$$t_\infty/t_* = n^{1/n-1}e^{1/n}\Gamma(1/n)$$

and we can see that at $n = 1$ value of t_∞/t_* is equal to e (the Neper's number) and with a fall of n this ratio increases.

M

Use of the Coulomb's Law for Description of Some Elastic-Plastic Systems at Cycling Loading

In some productions, for example, at deep boring the special device is used. It consists of the system of rings with conic polish sides (in Fig. M.1, a three such rings are shown). The main difficulty of the use of similar devices is finding such loading regimes at which the rings are sliding relatively to each other. Here the results of experimental-theoretical study of the apparatus is made (it was fulfilled by the author at Leningrad Polytechnic).

The experiments were provided on the Amsler's press and cycling machine CDM PU-100. At quasi-static loading the values of force P were taken from the press scale and deformation k – from two indicators. The simultaneous record of P and k at cycling loading was fulfilled by special electronic apparatus 90–16.2. To make the tests in liquid the device was situated in a cylinder.

The results of tests at quasi-static loading at different environment and α are given in Fig. M.2 (solid lines 1 for air at $\alpha = 0.331, 2$ – for lubricant with graphite at $\alpha = 0.331$, broken line 3 – for air at $\alpha = 0.262$, interrupted by points lines 4 – for oil at $\alpha = 0.262$ and 5 – for oil at $\alpha = 0.296$) and in Table M.1 under the same numbers. The character of graphs P(k) which have the form of triangles with upper sides responding to loading and vertical as well as lower straight lines – to unloading shows an opportunity of theoretical prediction of deformation characteristics of the device.

At the increase of upper or lower rings diameter D_b (Fig. M.1a) on value ΔD_b the system becomes shorter on $\Delta k_b = 0.5\Delta D_b \cot\alpha$. Similarly for other rings we have $\Delta\kappa_s = \Delta D_s \cot\alpha$. As a result the whole shortening of the system of n rings is

$$k = (\Delta D_b + (n-2)\Delta D_s)\cot\alpha. \quad \text{(M.1)}$$

When diameter D changes its value on Δ D the circumferential stresses appear. They can be replaced by radial forces q on the unit length of the perimeter (Fig. M.1b) as

$$q = 2EA\Delta D_i/D_i^2 (i = b, s) \quad \text{(M.2)}$$

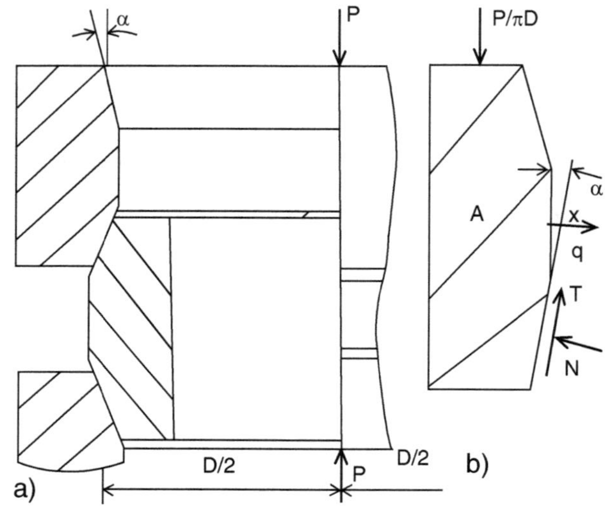

Fig. M.1. Scheme of ream of rings with their cross-section

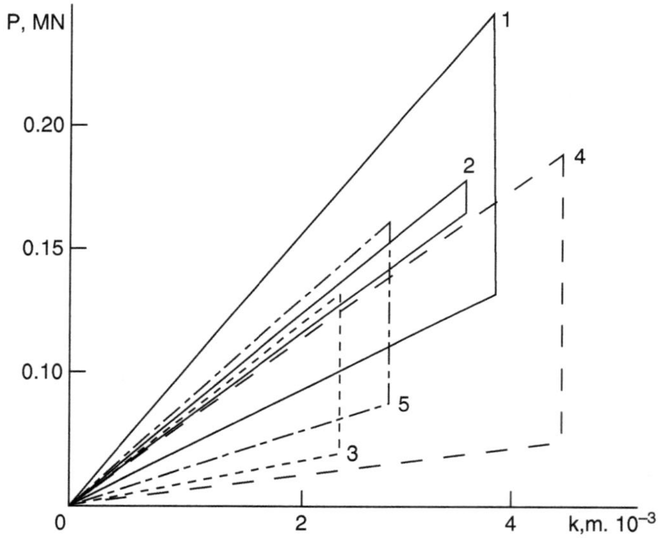

Fig. M.2. Results of experiments on quasi-static loading

where E is modulus of elasticity, A – area of cross-section of a ring (shaded in Fig. M.1). Making a sum of projections of forces on axis x in Fig. M.1a we have for upper or lower and middle rings respectively

$$\begin{aligned} -(P/\pi D_b)\cos\alpha + q_b\sin\alpha + f(q_b\cos\alpha + (P/\pi D_b)\sin\alpha) = 0, \\ -(P/\pi D_s)\cos\alpha + 0.5q_s\sin\alpha + f(0.5q_s\cos\alpha + (P/\pi D_s)\sin\alpha) = 0. \end{aligned} \tag{M.3}$$

Table M.1. Data of quasi-static loading

No	a (MN/m)	b (MN/m)	f	C (MN/m)
1	51.5	18.2	0.143	287
2	35	30	0.024	281
3	36	8	0.153	298
4	3ß	7.5	0.136	254
5	38	12.5	0.138	264

Here friction force T is found from the Coulomb's law $T = fN$, f – coefficient of friction and normal component N is computed according to the equilibrium equation.

Replacing in (M.3) values q_b, q_s by relations (M.2) and putting values $\Delta D_b, \Delta D_s$ into (M.1) we find after transformations

$$P = Ck\tan(\alpha + \pi f \operatorname{sign}(dk/dt))\tan\alpha, \tag{M.4}$$

where sign (dk/dt) reflects the difference of link between P, k at loading and unloading, π f – angle of friction and C is a rigidity factor which at $D_s = D_b$ is given by expression

$$C = pEA/(n - 1.5)D. \tag{M.5}$$

When a, b are tangents of angles between straight lines and axis k in Fig. M.2 at loading and unloading respectively, friction coefficient and rigidity factor can be found as

$$\begin{aligned} f^2 - f(a+b)/(a-b)\sin 2\alpha + 1 &= 0,\\ C^2 - 0.5C(a+b)\cot^2\alpha - 1 - ab\cot^2\alpha &= 0. \end{aligned} \tag{M.6}$$

In the solutions we must take before square root sign minus for the first relation and plus – for the second one.

Values a, b are also given in Table M.1 and we can see that f-values are near to their meanings in reference books and mean value of $C = 277\text{MN/m}$ coincides with that for tested reams at $n = 7, E = 2 \times 10^5\text{MN/m}^2, A = 3.15 \times 10^{-4}\text{m}^2, D = 0.13\,\text{m}$ according to (M.5).

The analysis of the test results on quasi-static loading gives an opportunity to suppose two work types of the device at cycling loading – with mutual slip of rings and without it (as piece of tube) with rigidity coefficients much more than follows from (M.4) as

$$C\tan(\alpha \pm \varphi)\tan\alpha.$$

The latter regime takes place if at the beginning of loading the minimum value of force in a cycle is more than force P^* responding to the beginning of rings

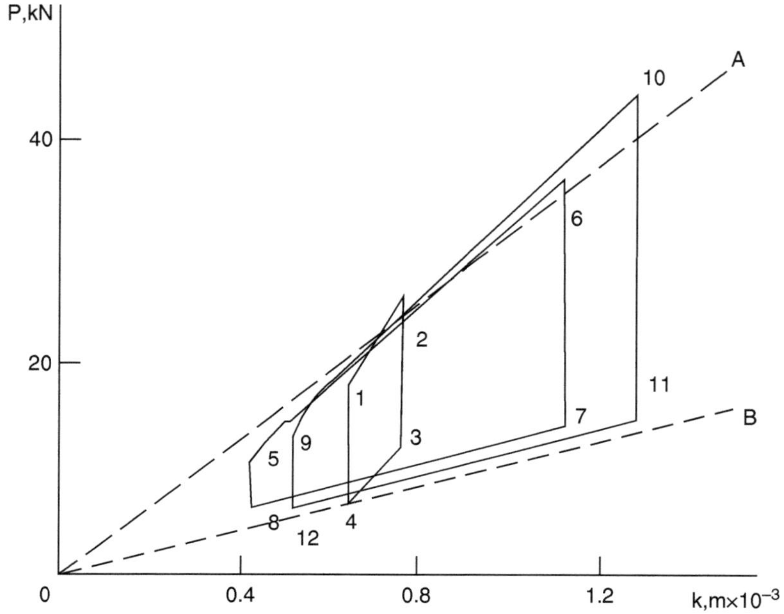

Fig. M.3. Function P(k) at cycling loading

Table M.2. Data of cycling loading

No of regime	Loop in figure	Frequency in Hz	(kN)		ψ	
			P_1	P_2	Test	Theory
I	1–4	21	25	5	0.72	0.68
II	5–8	21	35	5	0.21	0.30
III	9–12	21	55	5	0.22	0.18
	0-A-B-0	statics	any	0	0.08	0.08

slip at unloading. For P^* we have from (N.4)

$$P^* = P_1(\tan(\alpha + \varphi))/\tan(\alpha - \varphi) \tag{M.7}$$

where P_1 is maximum load in a cycle.

In Fig. M.3 as an example the test results for rings ream with $\alpha = 0.262$ in oil at frequency 21 Hz in regimes I–III of Table M.2 together with broken straight lines of loading OA and unloading OB at quasi-static deformation are given. It is seen some difference of form and position of hysteresis loops at quasi-static and quick loadings. It is the biggest for the regime with maximum of amplitude that may be explained by the divergence from the Coulomb's law with growth of a velocity of mutual displacement of sliding surfaces.

Now we find absorption coefficient ψ as the ratio of hysteresis loop area to potential energy $0.5\,Ck_a^2$ of the system where k_a is amplitude of the deformation. Theoretical values of ψ are computed according to following from (M.4) relation

$$\psi = 2((P_1 + P_2)0.5 \sin 2\alpha(P_1 - P_2)f)f \sin \alpha/((P_1 - P_2)0.5 \sin 2\alpha \\ -(P_1 + P_2))f/(1 - f^2 \tan^2 \alpha) \cos^3 \alpha.$$

where P_2 is minimum force in a cycle.

In conclusion we must notice that the difference between quick and slow deformations of the ream is not big. Using the data on quasi-static loading we can predict the character of the reams work including their dissipative losses at cycling loading.

The received results may be used for explanation of behavior of other elastic-plastic systems including some soils.

N

Investigation of Gas Penetration in Polymers and Rubbers

Since polymers and rubbers are often used as shells and membranes (see Sects 6.2.4 and 6.2.5) it is necessary to study gas penetration through them. Herewith such investigation is made (it was fulfilled by the author in Leningrad State University).

The process of gas penetration materials is often considered as successive phenomena of absorption and diffusion that go into direction of decreasing concentration gradient. Unsteady state of diffusion gas stream in a material is described by differential equation similar to (1.12)

$$\partial C/\partial t = D\partial^2 C/\partial x^2. \tag{N.1}$$

Here D is diffusion coefficient, C – concentration of gas, x – coordinate in the direction of its movement.

Steady state of diffusion stream subdues to relation

$$q = -D\partial C/\partial x \tag{N.2}$$

where q – quantity of gas penetrating through unit surface area of a material in unit time.

It is supposed that the gas absorption in polymers is described by the Henri's law

$$C = ßp. \tag{N.3}$$

Here ß – absorption coefficient, p – gas pressure.

With consideration of (N.3) expression (N.2) may be represented in form

$$Q = Dß\Delta pAt/h \tag{N.4}$$

where Q is a volume of gas that is diffused through a sample of material with thickness h and area A during time t at difference of pressures before and after the plate Δp.

If we define the penetration factor r as gas volume which goes through unit area of cross-section in second at unit gradient of pressure that is

$$r = Qh/At\Delta p \tag{N.5}$$

constants r, D, ß will be linked by relation

$$r = Dß. \tag{N.6}$$

Hence the gas penetration may be fully characterized by three constants – of diffusion D, absorption ß and penetration r. They can be found experimentally by different methods.

One of them allows to determine all the constants simultaneously from one test as follows.

For the plate at initial and border conditions

$$\begin{aligned} &C = C_o \text{ at } t = 0, 0 < x < h,\\ &C = C_1 \text{ at } x = 0, C = C_2 \text{ at } x = h \text{ for all } t \end{aligned}$$

we have the following solution of (N.1) /52/

$$\begin{aligned} C(x,t) = {} & C_1 + (C_2 - C_1)x/h + (2/\pi)\sum_{n=1}^{\infty}((C_2\cos\pi n - C_1)/n)\sin(n\pi x/h)\\ & \times \exp(-Dn^2\pi^2/h^2) + (4C_o/\pi)\sum_{m=0}^{\infty}(2m+1)^{-1}\sin((2m+1)\pi x/h)\\ & \times \exp(-D(2m+1)^2\pi^2 t/h^2). \end{aligned} \tag{N.7}$$

The gas stream through unit area of the plate cross-section into volume Vo is determined by expression

$$Vo\partial C/\partial t = -D\partial C/\partial x|_{x=h}. \tag{N.8}$$

Differentiating (N.7) by x and putting the result at $x = h$ into (N.8) we receive after integration from 0 to t following relation for gas volume Q which went through the plate at time t

$$\begin{aligned} Q = {} & ßA((p_1 - p_2)Dt/h + (h/\pi^2)(2\sum_{n=1}^{\infty}(p_1 - p_2\cos\pi n)(\cos n\pi)\\ & \times (1 - \exp(-Dn^2\pi^2 t/h^2))/n^2 - 4(p_o/\pi^2)\sum_{m=0}^{\infty}(\cos(2m+1)\pi)\\ & \times (1 - \exp((-D(2m+1)^2\pi^2 t/h^2))/(2m+1)^2) \end{aligned} \tag{N.9}$$

where p_o is the pressure at initial gas concentration C_o in the plate, p_1 and p_2 – pressures before and after the plate.

At big t after summation of series (N.9) becomes

$$Q = ßA((p_1 - p_2)Dt/h - (p_1 + 2p_2)h/6 + p_o h/2). \tag{N.10}$$

If $Q = 0$ we can find the "time of lagging" (at unsteady penetration) t_1 as

$$t_1 = h^2(p_1 + 2p_2 - 3p_o)/6D(p_1 - p_2). \tag{N.11}$$

At small p_1 and p_o comparatively to p_1 we receive

$$D = h^2/6t_1. \tag{N.12}$$

So, when we have experimental dependence Q(t) for unsteady and steady state of diffusion gas stream we can find with the help of relations (N.5), (N.12) and (N.6) constants r, D and ß.

For the experiment a plant was created which gives an opportunity to test material plates with thickness 0.1–1 cm and working area 110 cm^2. Duration of a test is determined by the time when the diffusion velocity becomes constant and it is equaled usually $2t_1$. The volume of a gas for computation is found according to the Clapeyron's relation

$$Q = 3.6H\Sigma\Delta Q_i/T \tag{N.13}$$

where ΔQ_i is the gas volume at considered time, H-atmospheric pressure in cm of a mercury pillar, T – temperature of test in °K. The diagrams Q(t) for 6 materials are given in Fig. N.1 and computed according to relations (N.5), (N.12), (N.6) constants – in Table N.1.

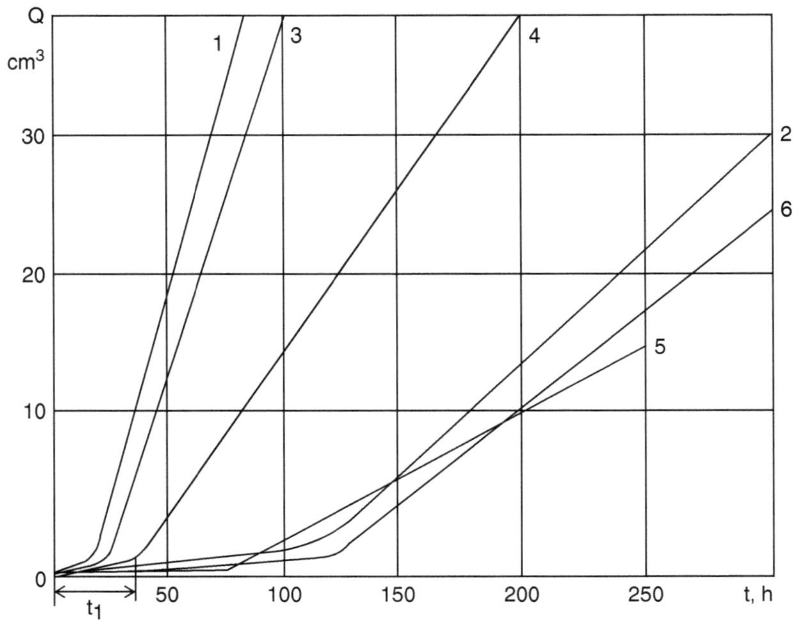

Fig. N.1. Diagrams Q(t)

Table N.1. Test data

No	Material	Δp (kn/cm^2)	Number of samples	h (cm)	r × 10^{10} (cm^4/sN)	D × 10^7 (cm^2/s)	ß × 10^3 (cm^2/N)
1	Rubber	0.5	4	0.22	5.76	1.64	3.54
2	Rubber	4	5	0.30	0.41	0.57	0.78
3	Polyvinilchlorid	0.5	5	0.23	4.26	1.29	3.33
4	Polypropilen	1	6	0.22	1.36	0.52	2.65
5	Polyethylene	0.5	5	0.31	0.08	0.92	1.36
6	Polycarbonate	2	5	0.27	0.47	0.22	1.93

The analysis of the results (here only a part of them is represented) shows that the penetration of the materials falls with a growth of pressure difference. It is also seen that the decrease of r takes place due to a decrease of D and ß. The change of penetration should be taken into account at a construction of rubber and polymer structures.

O

Fracture of Optimal Profile Rotating Disk

For this task we use the relation of rotation

$$\partial(\mathrm{h r \sigma_r})/\partial \mathrm{r} = \mathrm{h\sigma_\theta \delta v^2 r^2}$$

where δ – density of material, v – angular velocity. Putting here relation (6.14) and omitting multiplier 3/4 we receive in initial dimensions at $\sigma_r(a_o) = \sigma_r(b_o) = 0$ instead of (6.15)

$$\Omega^\mu \delta v^2 \int\limits_{a_o}^{b_o} h_o(r_o) r_o (r_o + u) dr_o = \int\limits_{a_o}^{b_o} h_o(r_o)(1 + u/r_o)^{-1-\alpha\mu} \ln^\mu(1 + u/r_o) dr_o. \tag{O.1}$$

We begin wth case $h_o = h_a a_o^2/r_o^2$ (solid line for $\rho = r_o/a_o$ and $h_a = h_o(a_o)$ in Fig. O.1) when at

$$\beta = b_o/a_o, \zeta = u/a_o \tag{O.2}$$

we obtain from (O.1)

$$\Omega^\mu \delta v^2 a_o^2 (\beta - 1 + \zeta \ln \beta) = \int\limits_1^\beta (1 + \zeta/\rho)^{-1-\alpha\mu} \ln^\mu (1 + \zeta/\rho) \, d\rho/\rho^2. \tag{O.3}$$

If $\alpha = 0$ we get from (O.3) after integration

$$\Omega^\mu \delta v^2 a_o^2 = (\ln^{1+\mu}(1 + \zeta) - \ln^{1+\mu}(1 + \zeta/\beta))/\zeta(1 + \mu)(\beta - 1 + \zeta \ln \beta). \tag{O.4}$$

Using criterion $d\zeta/dt \to \infty$ we have critical time $\Omega(t_*)$ for ζ_* as

$$\Omega(t_*)^\mu \delta v^2 a_o^2 = ((1 + \zeta_*)^{-1} \ln^\mu(1 + \zeta_*) - \beta^{-1}(1 + \zeta_*/\beta)^{-1} \times \ln^\mu(1 + \zeta_*/\beta))/(\beta - 1 + 2\zeta_{**} \ln \beta). \tag{O.5}$$

The solution of the system must be made by the attempt method when for real $\zeta = \zeta_*$ we obtain $\Omega(t_*) = \Omega(t)$. Diagrams $t_u = t_*(\beta)$ where t_u, t_* are

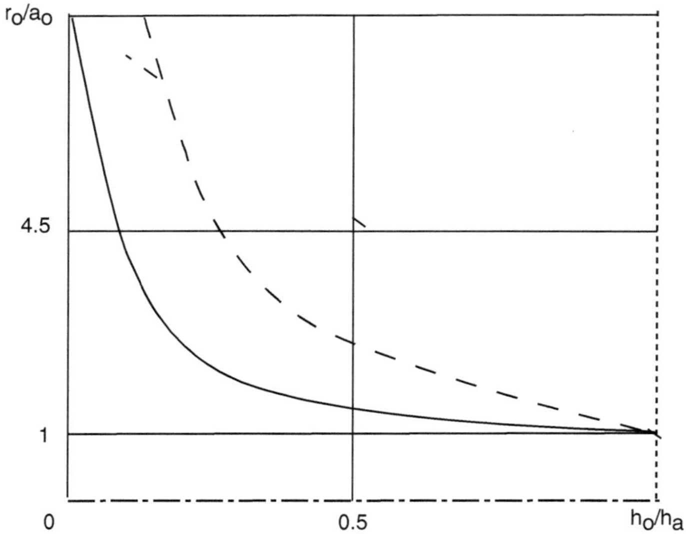

Fig. O.1. Profiles of disk

ultimate and critical times for $\mu = 1, \mu = 0.5$ are given in Figs. O.2 and O.3 by solid lines.

We investigate the role of parameter α for the case $m = 1$ when from (O.3) we have after integration

$$\Omega\delta v^2 a_o^2 = ((1+\zeta/\beta)^{-\alpha}(\ln(1+\zeta/\beta)+1/\alpha) - (1+\zeta)^{-\alpha} \times(\ln(1+\zeta)+1/\alpha))/\alpha(\zeta(\beta-1)+\zeta^2\ln\beta). \quad (O.6)$$

Using criterion $d\zeta/dt \to \infty$ we find

$$\Omega\delta v^2 a_o^2 = ((1+\zeta_*)^{-1-\alpha}\ln(1+\zeta_*) - \beta^{-1}(1+\zeta_*/\beta)^{-1-\alpha} \times\ln(1+\zeta_*/\beta))/(\beta-1+2\zeta_*\ln\beta). \quad (O.7)$$

From Fig. O.2 where at $\alpha = 1$ diagram $t_*(\beta)$ is given by pointed line we can see that with a growth of α the critical time diminishes.

In the same manner the case $h_o = h_a a_o/r_o$ (broken line in Fig. O.1) may be studied when from (O.1) at $\alpha = 0$ with consideretion of (O.2) we derive

$$\Omega^\mu \delta v^2 a_o^2 = \left(\int_1^\beta (1+\zeta/\rho)^{-1}\ln^\mu(1+\zeta/\rho)d\rho/\rho\right)/(\beta-1)(0.5(\beta+1)+\zeta). \quad (O.8)$$

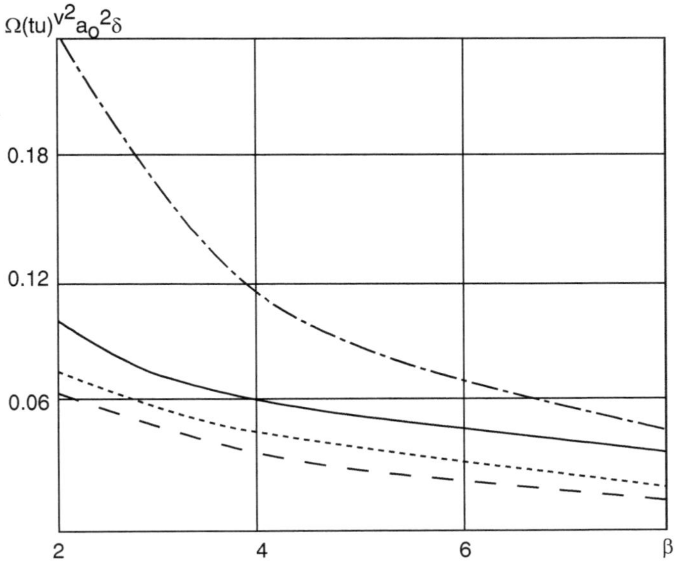

Fig. O.2. Dependence $t_u(\beta)$ for $m = 1$

Using criterion $d\zeta/dt \to \infty$ we compute

$$\Omega(t_*)^\mu \delta v^2 a_o^2 = ((1+\zeta_*)^{-1}\ln^\mu(1+\zeta_*) - (1+\zeta_*/\beta)^{-1} \times \ln^\mu(1+\zeta_*/\beta))/(\beta-1)\zeta_*. \tag{O.9}$$

The dependence of t_* on β for $\mu = 1$ and $\mu = 0.5$ is given in Figs. O.2 and O.3 by broken lines and we can see that they go lower consequent solid curves.

For the dick of constant thickness we have from (O.1), (O.2)

$$\Omega^\mu \delta v^2 a_o^2 = 6\left(\int_1^\beta (1+\zeta/\rho)^{1-\alpha\mu}\ln^\mu(1+\zeta/\rho)d\rho\right)/(\beta-1)(2(\beta^2+\beta+1)+3\zeta(\beta+1)). \tag{O.10}$$

According to criterion $d\zeta/dt \to \infty$ we compute

$$\Omega^\mu \delta v^2 a_o^2 = \int_1^\beta ((1+\zeta_*/\rho)^{-2-\alpha\mu}(\mu\ln^{\mu-1}(1+\zeta_*/\rho) - (1+\alpha\zeta_*\mu)\ln^\mu(1+\zeta_*/\rho))d\rho/\rho)/(\beta^2-1). \tag{O.11}$$

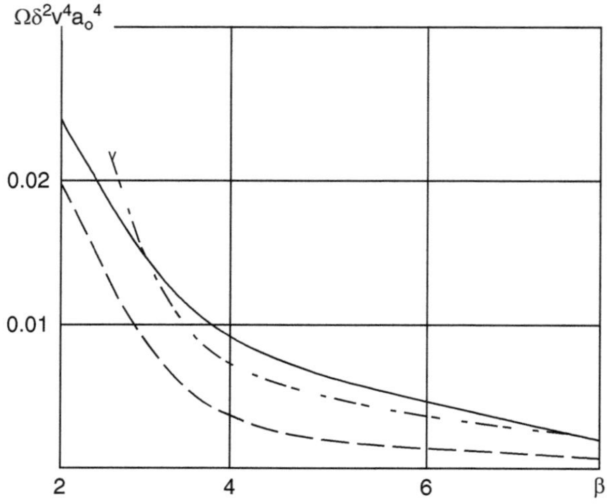

Fig. O.3. Dependence of ultimate time on β for $\mu = 0.5$

The diagrams according to (O.10), (O.11) for $\alpha = 0, \mu = 1$ and $\mu == 0.5$ (broken lines) go somewhat lower respective curves for the previous case and we can conclude that disk with h_o = constant is not an economical one.

To compare the above results with ones received by the Hoff's method (according to criterion $\zeta \to \infty$ for fracture time t_∞) we take from/17/

$$\Omega(\delta v^2 a_o{}^2)^m = \int_0^\infty (((\mu\beta+\zeta)(\beta+\zeta)^{-\mu} - (\mu+\zeta)(1+\zeta) - \mu/((1-\mu)\mu(A\zeta+B)^m)dt \quad (O.12)$$

where $A = (\beta^2 - 1)/2, B = (\beta^3 - 1)/3$.

L. Kachanov recommends to solve (O.12) approximately that is difficult to realize because of the infinity in the upper limit. However at $m = 2$ the rigorous result is possible in the form

$$\begin{aligned}\Omega_\infty(\delta a_{o2})^2 = {} & (16/A^2)(\ln((\sqrt{\beta}+1)^4/16\beta) + (2 - d_1/b_1)(d_1/b_1)\ln(\beta A/B) \\ & + (d_1^2/b_1 + d_2^2/b_2)/B + (d_2/b_2)(2 - d_2/b_2)\ln(A/B) \\ & + 2(d_1 d_2/b_1 b_2)(1 + \sqrt{\beta}A/B) + (d_1 d_2(b_1 + b_2)/b_1 b_2\end{aligned}$$

$$- 2(d_1+d_2))(\sin^{-1}((b_1 + b_2)/(b_1 - b_2))$$
$$- ((\sin^{-1}((b_1 + b_2)B + 2b_1b_2)/B(b_1 - b_2))/\sqrt{-b_1b_2}). \tag{O.13}$$

Here

$$b_1 = (\beta - 1)^2(\beta + 2)/6, \ b_2 = -(\beta - 1)^2(2\beta + 1)/6,$$
$$d_1 = -(\beta - 1)(\beta^2 + \beta + 4)/12, \ d_2 = -(\beta - 1)(4\beta^2 + \beta + 4)/12.$$

Diagram of (O.13) is given by interrupted by points curve in Fig. O.3 and it goes somewhat above the broken line.

The Hoff's method gives somewhat simpler solution for more economical profile $h_o = h_a a_o/r_o$ when at $\Omega(t) = B_1 t$ (the so-called steady creep) we derive

$$(du/dt)^\mu \int_{a_o}^{b_o} (r_o + u)^{-\mu-1} dr_o = \delta v^2 B_1 \int_{a_o}^{b_o} (r_o + u) dr_o. \tag{O.14}$$

For $\mu = 1, \mu = 0.5$ we compute respectively the fracture times

$$B_1 t_\infty \delta v^2 a_o^2 = 2(\beta - 1)^{-2} \ln((\beta + 1)^2/4\beta),$$
$$B_1 t_\infty (\delta v^2 a_o^2)^2 = 16(\beta - 1)^{-4}(\ln(\beta + 1)^2/4\beta + 2(2\sqrt{\beta}/(\beta + 1) - 1)).$$

The diagrams of these expressions are given in Figs. O.2 and O.3 by interrupted by points lines and we can see that they are above consequent broken curves for this profile. So the Hoff's method overestimates the strength of the structure. Since diagram for $\mu = 0.5$ is near to that at $h =$ constant we can suppose that at other μ the also near to each other. In conclusion we must underline that curved profiles not only economical but they give more ultimate time.

P

Strength of Anisotropic Tubes at Different Loadings. Construction of Potential Function

In order to receive the initial equations we use given in /38/ links between true σ_s $(s = x, y)$ and conditional σ_{so} stresses in form

$$\sigma_x = \sigma_{xo}(1 + 0.5n_o(\exp 2\varepsilon_y - 1)) \exp \varepsilon_x, \sigma_y = \sigma_{yo} \exp(\varepsilon_x + 2\varepsilon_y) \qquad (P.1)$$

where x, y – axial and circumferencial direction, $0 \leq n_o \leq \infty$ and the same for n. Putting these expressions into (2.39), (2.43) we find equations describing development of the tubes strains in time as

$$\varepsilon_x S_y = \varepsilon_y S_x, \exp(-\alpha\varepsilon_{eq})\varepsilon_y = \Omega(t)(\sigma_{xo})^{m-1} D^{m-1} S_y \qquad (P.2)$$

where $D = \sigma_{eq}(\exp \varepsilon_x)/\sigma_{xo}$ and values of S_x, S_y depend on the interposition of isotropy and loading planes in transversally isotropic tubes. For the cases when z,x,y are axes of symmetry they are respectively

$$S_x = 1 - 0.5n_o + n_o(0.5 - k) \exp 2\varepsilon_y,\ S_y = n_o(1 - 0.5k) \exp 2\varepsilon_y + k(0.5n_o - 1),$$
$$S_x = (1 - k)(2 - n_o),\ S_y = 0.5n_o(1 + k) \exp 2\varepsilon_y + (1 - k)(0.5n_o - 1),$$
$$S_x = 1 - 0.5n_o + n_o(k - 0.5) \exp 2\varepsilon_y,\ S_y = (1 - k)(1.5n_o \exp 2\varepsilon_y + 0.5n_o - 1).$$

Any stage of the deformation may be taken as allowable one but more convenient is the use of the infinite elongation rate when we get ultimate state naturally. Applying it to (P.2) as $d\varepsilon_y/dt \to \infty$ at n_o = constant we get equation for critical strain $\varepsilon_y = \varepsilon_*$ as following

$$(\varepsilon_*)^2 m(2 - n_o)n_o(1 - k^2) \exp 2\varepsilon_* - C_* S_{y*}\varepsilon_* + S_{y*}^2 = 0. \qquad (P.3)$$

If $R = (m - 1)dD/Dd\varepsilon_y + \alpha(d\varepsilon_{eq}/d\varepsilon_y)$ the functions $C(n_o, m, \varepsilon_y)$ for the cases above are respectively

$$C = RS_y + n_o((2 - k + m(0.5 - k))) \exp 2\varepsilon_y + m(1 - 0.5n_o),$$
$$C = RS_y + m(1 - k)(2 - n_o) + n_o(1 + k) \exp 2\varepsilon_y, \qquad (P.4)$$
$$C = RSy + mS_x + 3(1 - k)n_o \exp 2\varepsilon_y.$$

The critical time may be got from (P.2) at $\varepsilon_y = \varepsilon_*$.

Expressions (P.3), (P.4) do not contain t and hence can be used in the conditions when the influence of time is negligible and the limit state is appreciated in coordinates

$$Z_* = m\varepsilon_{e*},\ n = n_o(\exp 2\varepsilon_y)/(0.5n_o(\exp 2\varepsilon_y - 1) + 1) \tag{P.5}$$

where $Z = (d\varepsilon_e/d\sigma_e)\sigma_e$ – the subtangent to the hardening curve $\sigma_e(\varepsilon_e)$ (see Fig. 2.6). As a result we have instead of (P.3) equation for Z_*:

$$(Z_* \left(\sqrt{3}/2\sqrt{\Lambda_*}\right)^2 n_* (2 - n_*) \left(1 - k^2\right) - Z_* \left(\sqrt{3}/2\sqrt{\Lambda_*}\right) c_* + m = 0. \tag{P.6}$$

Here according to the criterion of the maximum shearing stress at $\sigma_{eq} = \sigma_y$ and $\sigma_{eq} = \sigma_x$ (upper and lower expressions respectively) we have at $\alpha = 0$

$$c = 1 + 2n(1 - m)x \begin{matrix}(1 - k + (2 - k)n)\\ (1 - 2nk + k^2),\end{matrix}$$

$$c = 2(1 - k + nk) + (m - 1)x \begin{matrix}(2k - 1 + 3n(1 - k))\\ (n^2 + 2(1 - n)(1 - k)),\end{matrix}$$

$$c = 1 + 2n(1 - k) + (m - 1)x \begin{matrix}(2k - 1 + 3n(1 - k))\\ (1 - 2n(1 - k)(n - 1)).\end{matrix}$$

From these relations we can see that at $m = 1$ the c-values do not depend on σ_{eq}. At other m near limit meanings of n $(0.\infty)$ it is necessary to use the conditions $\sigma_{eq} = \sigma_x$ and $\sigma_{eq} = \sigma_y$ respectively taking the least of two Z_*-values. In Fig. P.1 in polar coordinates Z_*, $a = \tan^{-1}\sqrt{3}n(2 - n)$ for k 0.5; 1, 0 – 1 the limit diagrams are constructed by solid, broken, interrupted by points and pointed lines respectively for the cases $m = 1$ and $m \to \infty$ (sign b and A means validity for all m)- The crosses correspond tests on nikel /40/ and confirms the theory. It allows to reinforce properly similar objects.

For an isotropic materials we have from (2.33) at $k_i = k_j = k_k = 0.5$ the initial basic equations (2.32). Here true stresses σ_x, σ_y are given by relations (P.1) and τ can be computed as $\tau = \tau_o \exp(\varepsilon_v - \varepsilon_y)$. The second law (P.2) is valid here with multiplier 0.25 in its right part and at $S_y = n_o(1 + 3\exp 2\varepsilon_y) - 2$.

According to criterion $d\varepsilon_y/dt \to \infty$ we receive expression (P.3) in which multiplier $1 - k^2$ must be replaced by 12 and for C we have

$$C = 6n_o \exp 2\varepsilon_y + 2m(2 - n_o) + ((m - 1)dD/Dd\varepsilon_y + \alpha(d\varepsilon_{eq}/d\varepsilon_y))S_x. \tag{P.7}$$

Similar to the anisotropic cases we can find the ultimate state in terms of Z_*, k and $f = \tau/\sigma_y$ where $f_o = 2f(\exp 2\varepsilon_y)/((2 - n)\exp 2\varepsilon_y + n)$. Putting this relations and (P.5) into expression for ε_* above we find

$$3n\,(2 - n)\,(Z_*)^2 - 2C_1\sqrt{1 - n + n^2 + 3f^2}\,Z_* + 4m\left(1 - n + n^2 + 3f^2\right) = 0 \tag{P.8}$$

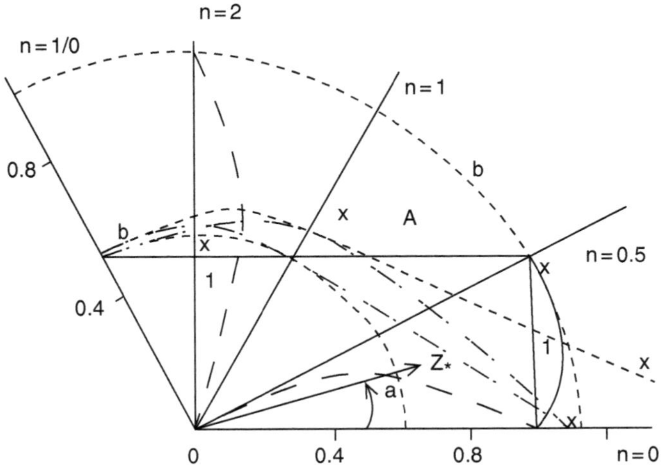

Fig. P.1. Diagram $Z_*(a)$ for z-axis of symmetry

where in the case $\sigma_{eq} = \sigma_e$

$$C_1 = 3n + m(2 - n) + (2n - 1)(3(m - 1)(n^2 - 2f^2)/2(1 - n + n^2 + 3f^2) + \alpha(d\varepsilon_{eq}/d\varepsilon_y)).$$

At $m \to \infty$, $\alpha = 0$ we find the ralations that were received earlier by Storakers on the base of the Drucker's postulate. In our case we can find other localization conditions at different σ_{eq}.

The tests on tubes at all three loading are rare. The case $f = 0$ is well confirmed by experiments on metals and polymers (see e.g. /14, 38, 39/). Very often the loading by axial force and torsion is used and for this option we have from (P.8)

$$Z_* = \sqrt{1 + 3f^2}(1 + (3(m - 1)f^2/(1 + 3f^2) + \alpha(d\varepsilon_{eq}/d\varepsilon_y))/2m)^{-1} \qquad \text{(P.9)}$$

In Fig. P.2 in coordinates $\tau\sqrt{3}$, $\sigma_x \equiv \sigma$ the lines 1, 4 $1/0 \equiv \infty$ are constructed at $\alpha = 0$. It is seen that curves 4, ∞ are near to each other and this justifies the use of classical criteria type /39/.and others which can be got from our results at $m \to \infty$ and different σ_{eq}.

It is also seen from Fig. P.2 that Z_* increases with a growth of τ Such phenomena are often occur in experiments e.g. for some steel /40/ (crosses in Fig. P.2). Supposing $\varepsilon_{eq} = \varepsilon_1$ (maximum elongation) and using notation $\tan\beta = \sqrt{3}\tau/\sigma$ we rewrite (P.2) as,

$$Z_* = ((1 + ((m - 1)\sin^2\beta + 0.5\alpha(1 + 3\sqrt{3}/\sqrt{3 + 4\tan^2\beta}))/2m)\cos\beta)^{-1}. \qquad \text{(P.10)}$$

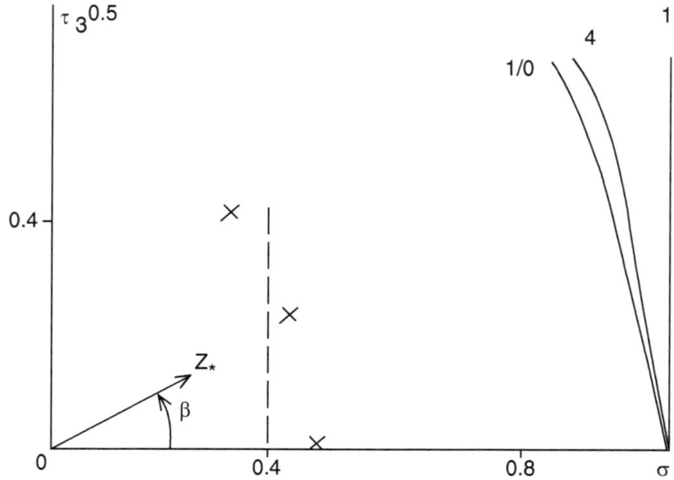

Fig. P.2. Comparison of theory and experiment for steel

Approximation of the first parts of creep curves gives m = 4 and the same procedure for the third parts shows that the damage factorα is about 6. The curve for these values is presented in Fig. P.2 by broken line and it is near the experimental data.

The potenial function (see (2.63), (2.64), (4.43)) is often used while solving practical problems (see e.g. /24/). At non-linear deformation when all invariants of strain tensor T_ε (2.13) depend on all invariants of T_σ these values are linked by relations /53/

$$\sigma_e \partial e/\partial\sigma_e - K\partial e/\partial K = \partial(\varepsilon_e \cos\omega)/\partial K, \tag{P.11}$$

$$\partial e/\partial\xi = -\partial(\varepsilon_e \sin\omega)/\partial K, \tag{P.12}$$

$$\begin{aligned}\partial\varepsilon_e/\partial\xi - K\tan\omega\partial\varepsilon_e/\partial K + \sigma_e \tan\omega\partial\varepsilon_e/\partial\sigma_e \\ = \varepsilon_e(K\partial\omega/\partial K - (1-\partial\omega/\partial\xi)\tan\omega - \sigma_e\partial\omega/\partial\sigma_e)\end{aligned} \tag{P.13}$$

where $e \equiv e_m$ (see (2.20)), σ_e, ε_e are given by relations (2.24), (2.25), $K = \sigma_m/\sigma_e$, (see (2.21)) phase of similarity ω is equal to the difference of angles ξ, ψ linked with the Lode's parameters μ_σ, μ_ε (see (2.34), (2.35)) by relations

$$\tan\xi = \mu_\sigma/\sqrt{3}, \; \tan\psi = \mu_\varepsilon/\sqrt{3}.$$

Here equations (P.11), (P.13) are used for the costruction of functions $\varepsilon_e(K, \sigma_e, \xi)$, $e(K, \sigma_e, \xi)$ for teflon-4 on the base of rheological law $\omega(K, \sigma_e, \xi)$ which

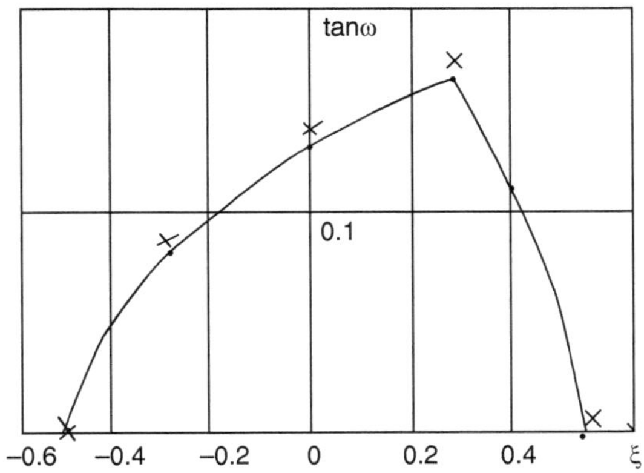

Fig. P.3. Function $\omega(\xi)$

Table P.1. Comparison of theoretical and test data

ξ	K	$\varepsilon/\varepsilon o$	
		Tests	Theory
$\pi/6$	−0.33	0.40	0.40
$\pi/12$	−0.17	0.45	0.51
0	0	0.55	0.68
$-\pi/12$	0.17	0.75	0.86
$-\pi/6$	0.33	1	1
$-\pi/12$	0.49	0.95	1.03
0	0.58	0.85	0.94
$\pi/12$	0.64	0.75	0.88
$\pi/6$	0.67	0.65	0.80

is taken in form

$$\tan\omega = \varphi(\sigma_e)(1 + \beta K)f(\xi). \quad (P.14)$$

Since expression for $\omega(K, \xi)$ has identical character we put $\varphi = 1$. The diagram $f(\xi)$ is given in Fig. P.3. The points and crosses refer to the cases of one compressive – one tensile (lines 1–4 in Table P.1) and two tensile (lines 5–9 in the Table) stresses respectively.

To construct function $\varepsilon_e(K, \sigma_e, \xi)$ we put (P.14) into (P.13) which gives

$$\partial\varepsilon_e/f\partial\xi - K(1+\beta K)\partial\varepsilon_e/\partial K + \sigma_e(1+\beta K)\partial\varepsilon_e/\partial\sigma_e$$
$$= \varepsilon_e((1+\beta K)^2 df/d\xi - 1 - (1+\beta K)^3 f^2)(1+(1+\beta K)^2 f^2)^{-1}. \quad \text{(P.15)}$$

To integrate (P.15) we find first function $\varepsilon_e(\sigma_e, \xi)$ at $K = 0$ and initial condition $\varepsilon_e = \Omega\sigma_e^m$ for $\xi = \xi_o$ which leads to expression

$$\varepsilon_e = \Omega\sqrt{1+f^2}\sigma_e^m F(\xi) \quad \text{(P.16)}$$

where

$$F(\xi)\exp\left(-(m+1)\int_{\xi}^{\xi} f d\xi\right)$$

In this case the solution is

$$\varepsilon_e = \Omega\sqrt{1+(1+\beta K)^2 f^2}\sigma_e^m(1+\beta K)^m F(\xi) \quad \text{(P.17)}$$

Similarity of curves $\varepsilon_e(K, \sigma_e, \xi)$ allows to compare values $\varepsilon_e/\varepsilon_o$ at certain σ_e, where $\varepsilon_o = \varepsilon_e$ for $\xi = \pi/6$ (simple tension). From columns 3, 4 of Table P.1 we can see the good agreement of the theory and the experiment at $\beta = 0.2$, $m = 3.5$. It is interesting to notice that at the same ξ value $\varepsilon_e/\varepsilon_o$ increases with a growth of K that may be explainet by the influence of normal stresses on slip planes (similar to the Coulomb's law).

Putting (P.14), (P.17) into (P.11), (P.12) we find after integration

$$e = \Omega\beta(1+\beta K)^m \sigma_e^m F(\xi) + G(\sigma_e, K).$$

With consideration of (P.14), (P.17) the last relation may be written in form

$$e = \beta\varepsilon_e \cos\omega + G(\sigma_m)$$

from which we can see that the volume change summarizes the parts proportional to mean ctress (2.21) and effective strain (2.21). The second one is positive and material becomes more friable that is confirmed by direct experiment.

Curriculum Vitae

ELSOUFIEV Serguey (ELSUF'EV Sergiy) was born in November 1935 in Omsk (Siberia, Russia). From February 1997 lives with his wife and a stepson constantly in Germany. Graduated with a distinction from the Hydrotechnical Faculty of Leningrad Polytechnic Institute in 1959 (the diploma project was presented in English). Worked as an engineer at building sites. In 1964 received his master's degree on the experimental bases of the Plasticity Theory. In 1985 defended his doctorate thesis in which he grounded a new scientific direction on strength computation of structures made of plastic metals, polymers, rubbers, soils etc. On the theme published 135 works, 29 of them in English and 15 his works are re-edited in USA. The book "Geomechanics" is edited in Odessa State University.

From 1964 to 1970 worked as a scientist in the Polymer Strength Laboratory at Leningrad State University, then as assistant, associate professor and professor in Leningrad Institutes (Polytechnic of Water transport). From 1987 till 1997 – as a professor, the Head of the Strength of Materials Department in Odessa State Academy of Civil Engineering and Architecture. Delivered lectures on the Strength of Materials and Structural Mechanics (3 years to foreign students in English), Theory of Elasticity, Plasticity and Creep, Applied Mechanics which combines very closely the Mechanism Theory, Machines Parts and Strength of Materials, and additionally in Leningrad and Odessa Universities, as well as in special institute – the Fracture Mechanics – a unique and original discipline. All the courses above are published (partly in English). Supervises the post-graduate students. For many years is a member of master and doctorate Councils on different branches of Mechanics. Is a member of GAMM (Society for Applied Mathematics and Mechanics).

Speaks Russian, English, German, Polish, understands French, Ukrainian. Has driving license. Healthy, is busy with sport (jogging, cycling, swimming). Neither drinks, nor smokes.

The contact address: Elsoufiev S., Vierhausstr.27, 44807 Bochum BRD (Germany).

Mobil: 4917665406935 E-Mail: Elsoufiev@gmx.de and elsoufie@et.rub.de

Index

MIX
Papier aus verantwortungsvollen Quellen
Paper from responsible sources
FSC® C105338

If you have any concerns about our products,
you can contact us on
ProductSafety@springernature.com

In case Publisher is established outside the EU,
the EU authorized representative is:
Springer Nature Customer Service Center GmbH
Europaplatz 3, 69115 Heidelberg, Germany

Printed by Libri Plureos GmbH
in Hamburg, Germany